JOURNAL OF CYBER
SECURITY AND MOBILITY

Volume 2, No. 3 - 4 (July - October 2013)

JOURNAL OF CYBER SECURITY AND MOBILITY

Aim
Journal of Cyber Security and Mobility provides an in-depth and holistic view of security and solutions from practical to theoretical aspects. It covers topics that are equally valuable for practitioners as well as those new in the field.

Scope
The journal covers security issues in cyber space and solutions thereof. As cyber space has moved towards the wireless/mobile world, issues in wireless/mobile communications will also be published. The publication will take a holistic view. Some example topics are: security in mobile networks, security and mobility optimization, cyber security, cloud security, Internet of Things (IoT) and machine-to-machine technologies.

JOURNAL OF CYBER SECURITY AND MOBILITY
COMMUNICATIONS
Volume 2, No. 3 - 4 (July - October 2013)

Published, sold and distributed by:
River Publishers
P.O. Box 1657
Algade 42
9000 Aalborg
Denmark

Tel.: +45369953197
www.riverpublishers.com

Journal of Cyber Security and Mobility is published four times a year.
Publication programme, 2014: Volume 2 (4 issues)

ISSN 2245-1439 (Print Version)
ISSN 2245-4578 (Online Version)
ISBN 978-87-93102-92-7 (this issue)

Foreword

Welcome to the volume 2 of the *Journal of Cyber Security and Mobility* that combines both issues 3 and 4 covering July issue and October issue of 2013, respectively. The current volume of the journal has 8 papers, four from each issue. These 8 papers cover topics such as physical layer security, access control, vulnerability prevention, timing, and synchronization issues in OFDM, light weight cryptographic trust and authentication mechanism to avoid emulation attacks, performance evaluation of 6LoWPAN and PANA in IEEE 802.15.4g mesh networks, and service ontology for Next Generation Service Overlay Network.

The first paper titled, "Adaptive Correction Algorithm for OFDM-IDMA Systems With Carrier Frequency OFFSET in A Fast Fading Multipath Channel ," by Balogun et al. investigates and analyzes the effect of CFO (Carrier Frequency Offset) on the performance of the OFDM-IDMA scheme in a fact fading multipath channel. This paper proposes an LMS-based adaptive synchronization algorithm to mitigate the degrading impact of carrier frequency offset errors on the OFDM-IDMA scheme.

The second paper titled, "Efficient Fine Grained Access Control for RFID Inter-Enterprise Systems," by Anggorojati, Prasad and Prasad, presents a novel access control model for an inter-enterprise RFID system. A system implementation of this model provides validates that the access control processing time is improved when this scheme is deployed.

In the third paper titled, "Dynamic AES – Extending the Lifetime," Henrik Tange and Birger Andersen propose a dynamic version of AES that can prevent some of the inherent vulnerabilities, namely side-channel attacks, related sub-key attacks and biclique attacks.

The fourth paper, titled, "Performance Evaluation of Beamspace MIMO Systems with Channel Estimation in Realistic Environments," by Maliatsos, Vasileiou and Kanatas analyzes the extended results for practical and realistic BS-MIMO transmission and reception schemes. It specifically focuses on channel estimation techniques for BS-MIMO systems with adaptive pattern reconfiguration.

Adaptation of the basic least-squares (LS) and minimum mean squared error (MMSE) estimators for the beamspace radio channels is performed and the algorithms are incorporated in an adaptive Singular Value Decomposition.

In the fifth paper, titled, "New Efficient Timing and Frequency Error Estimation in OFDM." Rathkanthiwar and Gandhi propose a new synchronization technique that uses training symbol to determine the timing and frequency synchronization error for OFDM systems. The proposed synchronization techniques result in better performance in terms of inter-symbol interference (ISI) and inter-carrier interference (ICI).

In the sixth paper titled, "Green Cooperative Web of Trust for Security in Cognitive Radio Networks," the authors Rohokale, Prasad and Prasad propose an energy efficient lightweight cryptographic Cooperative web of trust (CWoT) for the spectrum sensing in CRNs. They describe trust-based authentication and authorization mechanism for the opportunistic large array (OLA) structured Cognitive Radio Networks. They propose how Received Signal Strength (RSS) values obtained can be utilized to avoid primary user emulation attacks in CRNs.

The seventh paper titled, "Performance Evaluation on 6LoWPAN and PANA in IEEE 802.15.4g Mesh Networks ," authored by Ohba and Chasko evaluates the performance of 6LoWPAN and PANA over a mesh network based on mathematical analysis. These performance criteria include end-to-end IP packet error rate, mean end-to-end IP packet delay, PANA session failure rate and mean PANA session establishment delay. By means of performance analysis, the authors figure out a recommended PANA profile for IEEE 802.15.4g mesh networks.

In the eigth and final paper titled, "NGSON Service Composition Ontology," Schrage focuses on NGSON service ontology. The paper underlines the need for an OWL2 ontology that serves to provide machine understandable semantics. It defines the structure and required interfaces to other network entities, namely Software Defined Networks and Network Virtualization Functions.

We would like to thank the authors, anonymous reviewers, editorial board members, advisory board members, steering committee members, and the staff of River Publishers for their efforts towards publication of this issue of the journal. We hope the readers of this volume will enjoy the articles and get benefited from these timely topics. We look forward to contribution of technical papers from the readers in the area of cyber security and mobility.

Editors-in-Chief
Ashutosh Dutta, AT&T, USA
Ruby Lee, Princeton University, USA
Neeli Prasad, Aalborg University, Denmark

Adaptive Correction Algorithm for OFDM-IDMA Systems with Carrier Frequency OFFSET in a Fast Fading Multipath Channel

Muyiwa B. Balogun[1], Olutayo O. Oyerinde[2] and Stanley H. Mneney[1]

[1]*School of Engineering, Discipline of Electrical, Electronic and Computer Engineering, University of KwaZulu-Natal, Durban, 4041, South Africa, Email: 212561614@stu.ukzn.ac.za; Mneneys@ukzn.ac.za*
[2]*School of Electrical and Information Engineering, University of the Witwatersrand, Johannesburg, 2050, South Africa, Olutayo.Oyerinde@wits.ac.za*

Received 27 October 2013; Accepted 5 December 2013;
Publication 23 January 2014

Abstract

The Orthogonal Frequency Division Multiplexing-Interleave Division Multiple Access (OFDM-IDMA) scheme, which offers significant improvement on the performance of the conventional IDMA technique, has been in the forefront of recent mobile communication researches as it is expected to deliver a high quality, flexible and efficient high data-rate mobile transmission. Most papers on OFDM-IDMA scheme assume a system free of carrier frequency offset. However, the scheme is susceptible to synchronization errors and performance degradation because of the presence of OFDM, which is highly sensitive to carrier frequency offset (CFO) especially at the uplink. The effect of CFO on the performance of the scheme, in a slow fading multipath channel scenario, is therefore investigated, and analyzed. Also, the effect of CFO on the performance of the OFDM-IDMA scheme, in a fast fading multipath channel, which has not been hitherto reported in literature, is investigated and analyzed. An LMS-based adaptive synchronization algorithm is therefore employed to mitigate the degrading impact of carrier frequency offset errors on the OFDM-IDMA scheme. Simulation results clearly show that the presence of CFO degrades the performance of the system and that performance degradation

Journal of Cyber Security, Vol. 2 No. 3 & 4, 201–220.
doi: 10.13052/jcsm2245-1439.231

due to CFO, increases in a fast fading multipath channel in comparison with slow fading channel scenario. Furthermore, results show substantial improvement in the overall output of the system upon the application of the adaptive synchronization algorithm, which is implemented in both slow, and fast fading Rayleigh multipath channel scenarios.

Keywords: OFDM, IDMA, OFDM-IDMA, CFO, LMS.

1 Introduction

The need for better quality of service, improved capacity and high data-rate transmission, which has become nonnegotiable, have informed the continuous search for a reliable and efficient multicarrier scheme. The Code Division Multiple Access (CDMA) and the Orthogonal Frequency Division Multiplexing (OFDM) techniques rank high above other multiuser schemes due to their inherent advantages. The OFDM technique has particularly become difficult to ignore and almost indispensible because of its support for high data rate transmission and the ability to suppress ISI without much difficulty. Thus, OFDM has now become the bedrock of most recent multicarrier schemes in wireless communications. The combination of the CDMA and the OFDM technique to form a hybrid scheme of OFDM-CDMA has gained prominence and considered attractive due to the diversity and radio resource management flexibility offered. As studied in [1, 2, 3], there are various methods of combining the OFDM and the CDMA schemes, but the main idea behind the multicarrier CDMA hybrid scheme is to perform a spreading operation on transmitted signals which are then converted into parallel streams. The serial-to-parallel converted data are then modulated over different subcarriers, which are mutually orthogonal, and transmitted over the radio channel. The spreading code assigned to each user is to enable signal separation at the receiver. However, due to diverse level of fading and attenuation experienced by the transmitted signals, orthogonality is lost among subcarriers. This leads to Multiple Access Interference (MAI), causing high degradation in cellular performance, which becomes severe as the number of simultaneous users increases.

In an effort to address the MAI in multicarrier CDMA (MC-CDMA), the Multiuser Detection (MUD) technique was introduced. The priority of the MUD is to subtract interfering signals from the input signal of each user in the system. However, the MUD technique utilized in MC-CDMA comes with associated complexities and high cost [4, 5, 6]. Various MUD techniques have been proposed to address the high complexity of the MUD technique, but with little success. The complexity of the MUD tends to increase exponentially as the number of active subscribers increases. Recent studies have explored the possibility of the artificial neural network for multiuser detection [7] but these techniques tend to compromise system performance and efficiency for reduced complexity. The MUD challenge in MC-CDMA therefore remains and there has been a continuous search for an efficient and reliable scheme with low complexity.

To this end, a new multiuser scheme was recently proposed by Li Ping called the Interleave Division Multiple Access (IDMA) [8]. This scheme employs a simple low cost chip-by-chip iterative method for its multiuser detection. The IDMA scheme, which offers a lower system complexity compared to MC-CDMA [9], relies solely on interleaving as the only means of identifying signals from active users in the system. In a bid to achieve an improved cellular performance of the IDMA over multipath, Mahafeno in 2006 proposed an OFDM-based hybrid scheme called the OFDM-IDMA scheme [10]. The newly proposed multicarrier IDMA (MC-IDMA) scheme therefore combats ISI and MAI effectively over multipath with low complexity. The multicarrier scheme ensures a better cellular performance, high diversity order, and spectral efficiency compared to the MC-CDMA scheme. Thus, the scheme combines all the inherent advantages of the conventional IDMA and the OFDM technique. The associated MUD is of low cost and low complexity per user, which is independent of the number of simultaneous users in the system [11, 12].

Major studies on the OFDM-IDMA scheme focus only on the implementation of the scheme in a perfect scenario, assuming that there are no synchronization errors in the system. This is not obtainable in practice. The scheme becomes susceptible to synchronization errors because of the presence of the OFDM technique, which is highly sensitive to carrier frequency offsets (CFO) especially at the uplink, where different users are transmitting asynchronously, experiencing different levels of channel fading and delays. The CFO is caused mainly by Doppler Shifts or because of instabilities in the local oscillators [13]. CFO causes Inter-channel Interference (ICI) and loss of orthogonality among the sub-carriers, which in turn leads to performance degradation.

This paper, therefore, investigates and analyzes the impact of CFO on the general performance of the OFDM-IDMA scheme. Also, the performance of the system in a fast fading multipath channel is investigated and as well analyzed. As an extension of the performance analysis of the OFDM-IDMA scheme with synchronization errors, carried out in [14], an LMS-based adaptive synchronization algorithm is then introduced to combat the degrading impact of synchronization errors on the OFDM-IDMA system. Simulation results show that the performance of the system degrades as the CFO increases and the system also experiences degradation as the velocity increases in the fast fading multipath channel. An improved system output is however obtained upon the application of the LMS-based adaptive synchronization algorithm.

The rest of the paper is organized as follows: Section II describes the system model of the OFDM-IDMA scheme in the presence of CFO. Section III describes the system performance in the presence of CFO. Section IV presents the LMS-based adaptive synchronization algorithm. Section V discuses simulation results and finally Section VI gives the conclusion with a summary of the major results.

2 System Model

The OFDM technique [15,16], which involves the splitting of high-rate data streams into a large number of lower-rate data streams, which are transmitted, while maintaining orthogonality, across multiple narrowband sub-carriers, solves the problem of inter-symbol Interference (ISI), which is encountered while transmitting high-rate data streams across multipath channels. As long as orthogonality is maintained, there will be no interference between sub-carriers and this will enable the receiver to separate signals carried by each sub-carrier. Unlike the conventional Frequency Division Multiplexing (FDM) scheme, the spectra of the different modulated sub-carriers overlap in OFDM as seen in Fig. 1(b). This makes OFDM an appropriate scheme for optimum and efficient use of valuable spectrum. Fig. 2 illustrates in block diagram the adaptation of the IEEE standard 802.11a [17] for a baseband OFDM transceiver.

The IDMA technique is a type of multi-user technique where interleavers serve as the sole means of user separation. The interleavers, which are randomly generated, must be different for each user. Adjacent coded data, called "chips," are uncorrelated in this scheme as the interleavers separate the coded sequences, ensuring the simple implementation of the low-cost chip-by-chip detection algorithm [18, 19]. The OFDM transceiver structure in Fig. 2 is incorporated into the IDMA transceiver as shown in fig. 3. Considering a multicarrier IDMA system with K users transmitting simultaneously, for anyone of the users denoted by k, the input data array is first encoded using a Forward Error Correlation (FEC) code [20]. The coded sequence obtained is then interleaved by a randomly generated interleaver to give $x_k(n)$. The resulting signal is transmitted over the channel to the receiver. The received signal can be expressed as

$$r(n) = \sum_{k=1}^{k} x_k(n) h_k(n) + q(n) \tag{1}$$

$$= x_k(n) h_k(n) + \zeta_k(n) \tag{2}$$

$$\zeta_k(n) = \sum_{k' \neq k} x_{k'}(n) h_{k'}(n) + q(n) \tag{3}$$

where $h_k(n)$, which is assumed to be known at the receiver, is the fading channel coefficient for user k, $q(n)$ is the additive white Gaussian noise and the multi-user interference with respect to user k is denoted by $\zeta_k(n)$.

The receiver structure consists of the Elementary Signal Estimator (ESE) and a posterior probability (APP) decoders [20] for each user k. The mode of operation of the ESE and the decoders (DECs) is iterative [20] [21], deinterleaving and interleaving process also occurs iteratively between them. The ESE carries out a coarse chip-by-chip detection to roughly subtract the interference among the concurrent users in the multicarrier IDMA system. The chip-by-chip detection is of a very low computational cost and complexity [22]. The outputs of the ESE are the estimated probabilities of the transmitted signals, which are organized sequentially according to the simultaneous users and are fed to the APP decoders [22]. The mode of operation of the ESE and DECs is iterative, where extrinsic information is processed in a turbo-like mode between them. At the last iteration, the APP decoders give the hard decisions based on the refined estimations by the ESE and the logarithm likelihood ratio (LLR) estimate of the ESE and the APP decoders is obtained as [23]

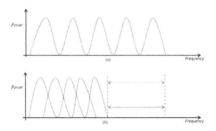

Figure 1 Illustration of the spectrum-saving concept of the OFDM technique (b) compared with the regular FDM scheme (a)

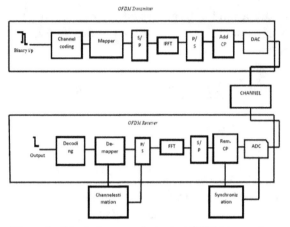

Figure 2　Block diagram of a typical OFDM transceiver

$$p\left(x_k\left(n\right)\right) \equiv log\left[\frac{P_b(x_k\left(n\right) = +1)}{P_b(x_k\left(n\right) = -1)}\right] \tag{4}$$

The outputs of both the ESE and the DECs above are probabilities of the values of the transmitted radio signals at the receiver [20], which will henceforth be stated as P_{ESE} (x_k(n) and P_{DEC}(x_k(n)) depending on whether they are emanating from the ESE or the APP decoders at the receiver. Therefore, considering the fading channel represented by the coefficient h, the ESE employs the received signal $r(n)$ and the LLR for its operation, so that the resulting output is obtained as [23]

$$log\left[\frac{P_b(x_k\left(n\right) = +1|r\left(n\right), h)}{P_b(x_k\left(n\right) = -1|r\left(n\right), h)}\right] = log\left[\frac{P_b(x_k\left(n\right) = +1), h}{P_b(x_k\left(n\right) = -1), h}\right]$$

$$+p_{ESE}\left(x_k\left(n\right)\right), \tag{5}$$

where the first part of the equation can be expressed as

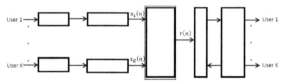

Figure 3　Block diagram of an OFDM-IDMA transceiver

$$e_{ESE}\left(x_k\left(n\right)\right) = \log\left[\frac{P_b(x_k\left(n\right) = +1), h}{P_b(x_k\left(n\right) = -1), h}\right], \tag{6}$$

which is the extrinsic LLR about the transmitted signal $x_k(n)$ based on the characteristics of the fading channel and the *a priori* information of concurrent users in the system [23].

Considering the ESE chip-by-chip detection in a quasi-static multipath channel where the BPSK signaling is used and the transmitted signal $x_k(n)$ is treated as a random variable, $e_{ESE}(x_k(n))$ is used to coarsely update the *a priori* LLR $p(x_k(n))$, which is obtained from (4) as [23]

$$E\left(x_k\left(n\right)\right) = \left[\frac{exp\left(p_{ESE}(x_k\left(n\right))\right) - 1)}{exp\left(p_{ESE}\left(x_k\left(n\right)\right)\right) + 1}\right] = tanh\left(p_{ESE}(x_k\left(n\right))/2\right), \tag{7}$$

$$Var\left(x_k\left(n\right)\right) = 1 - \left(E\left(x_k\left(n\right)\right)\right)^2, \tag{8}$$

where $E(x_k(n))$ and $Var(x_k(n))$ are the mean and variance of the transmitted signal $x_k(n)$ respectively. Using the central limit theorem, the interference $\zeta_k(n)$ in (3) can be estimated by a zero-mean Gaussian variable with variance σ^2, given as [23]

$$E\left(\zeta_k\left(n\right)\right) = \sum_{k'\neq k}^{K} h_{k'}\left(n\right) E\left(x_{k'}\left(n\right)\right), \tag{9}$$

$$Var\left(\zeta_k\left(n\right)\right) = \sum_{k'\neq k}^{K} |h_{k'}\left(n\right)|^2 Var\left(x_{k'}\left(n\right)\right) + \sigma^2. \tag{10}$$

Applying the Gaussian estimation to the received signal in (2), the output of the ESE in (3) is obtained as [23]

$$e_{ESE}\left(x_k\left(n\right)\right) = 2h_k\left(n\right).\frac{r(n) - E(\zeta_k(n))}{Var(\zeta_k(n))}$$

$$= 2h_k\left(n\right).\frac{r\left(n\right) - E\left(r\left(n\right)\right) + h_k\left(n\right) E\left(x_k\left(n\right)\right)}{Var\left(r\left(n\right)\right) - |h_k\left(n\right)|^2.Var\left(x_k\left(n\right)\right)}. \tag{11}$$

The estimated mean and variance of the received signal based on (1) is therefore obtained as [23]

$$E\left(r\left(n\right)\right) = \sum_{k'\neq 1}^{K} h_{k'}\left(n\right) E\left(x_{k'}\left(n\right)\right), \tag{12}$$

$$Var\left(r\left(n\right)\right) = \sum_{k' \neq 1}^{K} |h_{k'}\left(n\right)|^2 \, Var\left(x_{k'}\left(n\right)\right) + \sigma^2. \tag{13}$$

In [25, 26], various methods for carrier frequency offset estimation have been presented and these methods of CFO estimation can be put to use in the OFDM-IDMA system. Therefore, the carrier frequency offset is assumed to be known for users in this OFDM-IDMA model.

3 System Performance Analysis

The expression in (*1*) represents only a perfect scenario where the CFO is zero. Considering, therefore, the effect of CFO on the received baseband signal, (*1*) can be rewritten as

$$r_c\left(n\right) = \sum_{k=1}^{K} x_k\left(n\right) h_k\left(n\right) e^{j2\pi\varepsilon_k n/N} + q\left(n\right), \tag{14}$$

where n represents the sub-carrier index, N is the number of sub-carriers and ϵ_k, $\epsilon_k \ll 0.5$ [24], represents the normalized CFO. After DFT processing in the presence of CFO, equation (*14*) becomes:

$$R_c\left(m\right) = \sum_{n=0}^{N-1} r_c\left(n\right) e^{-j2\pi\varepsilon_k m/N} \tag{15}$$

$$= X_k\left(m\right) H_k\left(m\right) + \sum_{k' \neq k} X_{k'}\left(m\right) H_{k'}\left(m\right) + \mu_k\left(m\right) + Q(m) \tag{16}$$

The presence of CFO in the OFDM-IDMA system and its impact on the performance of the system can be inferred from (14)–(16), which are the modifications of (1)–(3). The second part of (16), which gives the combined interference due to the introduced CFO and the multiuser interference, can be represented by ζ'_k (m) given as:

$$\zeta'_k\left(m\right) = \sum_{k' \neq k} X_{k'}\left(m\right) H_{k'}\left(m\right) + \mu_k\left(m\right) + Q(m) \tag{17}$$

where Q(m) represents a Gaussian random variable which can be expressed as [27]:

$$Q\left(m\right) = \sum_{n=0}^{N-1} q\left(n\right) e^{-j2\pi n(m-\varepsilon_k)/N}, \tag{18}$$

and $\mu_k\left(m\right)$ is the interference due to CFO between user k given as ϵ_k and other user k' given as $\epsilon_{k'}$. This can be expressed following [28] as:

$$\mu_{\text{k}}\left(\text{m}\right) = \sum_{n=0}^{N-1} e^{j2\pi n(\varepsilon_{k'} - \varepsilon_k)/N}. \tag{19}$$

The interference in (19) can be further expressed as [27]

$$\mu_k\left(m\right) = \sum_{n=0}^{N-1} \frac{\sin(\pi \varepsilon_{k'})}{N \sin(\pi(\varepsilon_{k'} - \varepsilon_k)/N)} \cdot e^{j\pi(\varepsilon_{k'} - \varepsilon_k)(N-1)/N}. \tag{20}$$

As shown in Fig. 3, the chip-by-chip detection is carried out involving the ESE function, and the DECs perform the decoding operation using the output of the ESE as the input. Now considering the scenario where detection takes place in a multipath channel, with BPSK signaling assumed, the interference mean and variance based on [29] are given by

$$E\left(\zeta'_k\left(m\right)\right) = E\left(R_c\left(m\right)\right) - H_{k'}\left(m\right)\mu_k\left(m\right)E(X_{k'}(m)) \tag{21}$$

$$Var\left(\zeta'_k\left(m\right)\right) = Var\left(R_c\left(m\right)\right) - |H_{k'}\left(m\right)\mu_k\left(m\right)|^2 Var\left(X_{k'}\left(m\right)\right) \tag{22}$$

From the equations above, the output of the elementary signal operator in the presence of CFO, in a multipath channel, based on the extrinsic log-likelihood ratios (LLRs) generation [29], is represented by

$$e'_{ESE}\left(X_k\left(m\right)\right) = 2H_k(m)\frac{R_c\left(m\right) - E(\zeta'_k\left(m\right))}{Var(\zeta'_k\left(m\right))} \tag{23}$$

The above equations therefore represent the expressions for the OFDM-IDMA system model in the presence of carrier frequency offset. The model used combines the signal for all users at the receiver. The established presence of CFO and its degrading impact on the OFDM-IDMA system in a multipath channel should therefore necessitate the need for an effective correction and synchronization technique to improve the performance and the efficiency of the multiuser scheme.

4 The LMS-BASED Adaptive Synchronization Algorithm

The carrier frequency offset of a particular user can be compensated coarsely in the time domain of the system. But the main challenge is the residual CFOs due to multiple active users in the system, which results in interchannel interference (ICI). Due to system configuration of the OFDM-IDMA scheme, the mitigation of the effect of the residual CFOs is executed after the fast Fourier transforms (FFT) process, in the frequency domain. The proposed LMS-based adaptive synchronization algorithm is utilized to compensate the residual CFOs due to simultaneous users in the system. The characteristics of the LMS algorithm have been broadly studied and the most striking feature of the LMS-based algorithm is its simplicity [30]. The coefficient vector of the adaptive algorithm is updated by the expression [31]

$$L\left(n+1\right) = L\left(n\right) + 2\mu \times e\left(n\right) x_k\left(n\right), \tag{24}$$

where $L(n)$ is the coefficient vector, μ is the step-size which regulates the convergence speed of the synchronization algorithm and $x_k(n)$ is the modulated input signal utilized in the implementation of the algorithm. The update error signal $e(n)$ is expressed as

$$e\left(n\right) = r\left(n\right) - r_c\left(n\right), \tag{25}$$

where $r(n)$ represents the desired signal and $r_c(n)$ is the received signal in the presence of CFOs. The error signal is therefore used to update the carrier frequency-tracking loop, and the expected output is represented as [31]

$$y\left(n\right) = L\left(n\right) x_k\left(n\right), \tag{26}$$

Thus, with adequate knowledge of the input data by the receiver and the update error signal obtained, the impact of the CFOs can successfully be corrected for the users in the OFDM-IDMA system.

The non-recursive form of (24) is expressed as

$$L\left(n\right) = L\left(0\right) + 2\mu \sum\nolimits_{i\,=\,0}^{N-1} e\left(i\right) x_k\left(n\right). \tag{27}$$

From (16) and (17), initializing L(0) = 0, $y(n)$ is derived as

$$y\left(n\right) = 2\mu \sum\nolimits_{i\,=\,0}^{N-1} e(i) x'_k\left(n\right), x_k\left(n\right) \tag{28}$$

It is essential however, to state that the derived synchronization method is sensitive to the step-size as it is the case for most adaptive algorithms. Hence, the step-size μ is carefully determined [32] to obtain an efficient result and a reasonable convergence time.

5 Simulation Results and Discussion

This section presents the computer simulations carried out to analyze and substantiate the performance of the OFDM-IDMA system with CFO in fading multipath channel scenario. The OFDM-IDMA system model used has four users for all instances with input data length 32, spreading length 4. The number of sub-carriers is N = 128, the number of samples in the guard interval is set at 7 and the number of iteration is 4. In practice, the number of sub-carriers N is varied to provide variable data rates for different users. However, the same number of sub-carriers is assumed for all users in the simulation for convenience. The QPSK modulation technique is used assuming operating carrier frequency of 2GHz with sampling period of 0.5μ s. All simulation results are presented based on the bit error rate (BER) [33] performance of the system in a Rayleigh fading multipath channel of M =16 paths with normalized Doppler frequencies of fDn = 0.0136, fDn = 0.1085 and fDn = 0.1808 corresponding to mobile speeds of v = 15km/h, v = 120km/h and v = 200km/h respectively. The general performance of the system is investigated for a constant slow fading scenario with mobile speed v = 15km/h and then at increasing mobile speeds v = 120km/h and v = 200km/h.

In Fig. 4, the general performance of an OFDM-IDMA system in the presence of CFOs is shown. Various values of CFOs are varied from zero (when there is no synchronization error in the system) to CFO = 0.18, with the mobile speed constant at 15km/h in the fading Rayleigh multipath channel. It can be seen from the plot that at small value of CFO (i.e. 0.02), the system performance is not significantly affected. But as the values increase, performance degradation increases as well.

Figure 5 shows the impact of increasing mobile speed on the OFDM-IDMA system in a fast fading Rayleigh multipath channel. The performance of the system is affected as the mobile speed is increased. It can be seen that at CFO = 0.1, with the mobile speed also increasing in a fast fading Rayleigh channel scenario, as shown in figure 5, there is significant degradation in the system performance. This means further increase in the value of CFO in the fast fading multipath channel scenario, will only worsen the performance of the system.

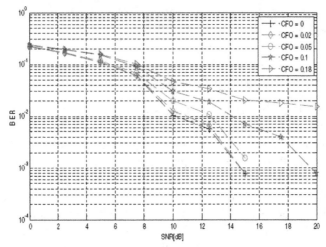

Figure 4 General BER performance of the OFDM-IDMA system model with increasing CFOs

In Fig. 6, the bit error rate performance of the OFDM-IDMA system with the application of the proposed LMS-based adaptive synchronization algorithm is presented. The plot shows the impact of the proposed algorithm on the system model in the presence of CFOs of 0.05 and 0.1. There is an appreciable reduction and significant

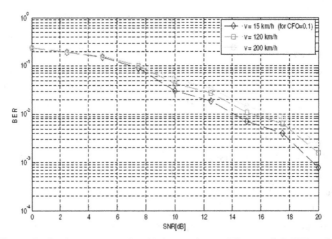

Figure 5 Performance result with increasing mobile speed at CFO = 0.1

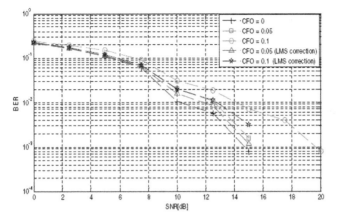

Figure 6 The impact of the proposed KLMS synchronization algorithm on the OFDM-IDMA system model with carrier frequency offsets 0.05 and 0.1

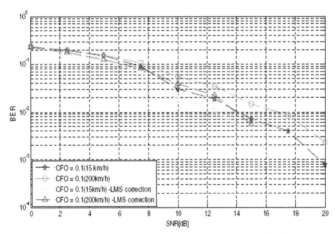

Figure 7 BER performance of the OFDM-IDMA system model with the proposed algorithm in both slow fading and fast fading multipath channel of mobile speeds 15 km/h and 200 km/h respectively

mitigation of the carrier frequency offset errors upon the application of the proposed algorithm. This signifies an effective reduction in the ICI due to the residual CFOs of concurrent users in the multicarrier IDMA system.

Furthermore, the performance of the OFDM-IDMA system model in a multipath channel of varying mobile speed is demonstrated in Fig. 7. The carrier frequency offset value is fixed while the mobile speed is varied at 15 km/h, and 200 km/h. The overall bit-error rate performance of the system degrades as the mobile speed increases. The introduction of the proposed algorithm, however, is able to mitigate the overall impact of carrier frequency offset error on the system, even in a fast fading multipath channel scenario.

6 Conclusion

The performance of the OFDM-IDMA system has been investigated and analyzed in a fast-fading Rayleigh multipath channel, in the presence of CFO. Simulation results clearly show that CFOs impact the system performance adversely in contrast to many research works on the OFDM-IDMA where it is assumed that CFOs have no influence on the overall performance of the system. Also, there is further degradation of the system in a fast fading multipath channel. An adaptive synchronization algorithm, which is LMS-based, has been presented to mitigate the impact of carrier frequency offset errors on the recently proposed OFDM-IDMA scheme. The algorithm focuses on the reduction of the ICI due to the residual CFOs from multiple active users in the system. The proposed algorithm was carried out in the presence of high carrier frequency offset values for clear demonstration of its efficiency. Simulation results also show an appreciable reduction and significant mitigation of the carrier frequency offset errors upon the application of the proposed algorithm in a fast fading Rayleigh multipath channel.

Reference

[1] R. L. Pickholtz, L. B. Milstein and D. L. Schilling. Spread spectrum for mobile communications, IEEE Trans. Vehicular Technology, vol. 40, pp. 313–322, May 1991.
[2] R. Prasad and S. Hara. An overview of multi-carrier CDMA, Int. Symp. IEEE Spread Spectrum Techniques and Applications Proceedings, vol. 1, pp. 107–114, Sept. 1996.
[3] A. McCormick, E. Al-Susa. Multicarrier CDMA for future genera-tion mobile communications, IEE Electronics & Comm.," Engineering, Vol. 14, Issue 2, Page(s): 52–60, April 2002.
[4] S. Moshavi. Multi-user Detection for DS-CDMA Communications. *IEEE Commun. Mag.*, vol. 34, pp.124–36, Oct. 1996.

[5] D. W. Matolak, V. Deepak, and F. A. Alder. Performance of Multitone and Multicarrier DS-SS in the presence of imperfect phase synchronization, MILCOM 2002, vol. 2, pp. 1002–1006, Oct 2002.

[6] K. Rasadurai and N. Kumaratharan. Performance enhancement of MCCDMA system through turbo multi-user detection, Computer comm. and Infomatics (ICCI) 2012, pp. 1–7, 2012.

[7] F. Corlier and F. Nouvel. Unsupervised neural network for Multi-user detection in MC-CDMA systems," IEEE Int. Conf. on Personal wireless comm., pp. 255–259, 2002.

[8] Li Ping, K. Y.Wu, L. H Liu and W. K. Leung, A simple unified approach to nearly optimal multiuser detection and space-time coding, Information TheoryWorkshop, ITW'2002, pp. 53–56, October 2002.

[9] Li Ping. Interleave-division multiple access and chip-by-chip iterative multi-user detection, *IEEE Commun. Mag.*, vol. 43, no. 6, pp. S19–S23, June 2005.

[10] I. Mahafeno, C. Langlais, and C. Jego. OFDM-IDM Aversus IDMA with ISI cancellation for quasi-static Rayleigh fading multipath channels, in Proc. 4th Int. Symp. on Turbo Codes & Related Topics, Munich, Germany, Apr. 3–7, 2006.

[11] K. Kusume, G. Bauch, W. Utschick. IDMA vs. CDMA: analysis and comparison of two multiple access schemes, IEEE Trans. Wireless Commun., vol. 11, pp. 78–87, Jan. 2012.

[12] Li Ping, Qinghua Guo and Jun Tong. The OFDM-IDMA Approach to Wireless Communication System, IEEE Wireless Communication, pp.18–24, June 2007.

[13] M. Morelli, A. D'Andrea, and U. Mengali, Feedback frequency synchronization for OFDM applications, IEEE Communication Letter, vol. 5, pp. 134–136, Jan. 2001.

[14] M. B Balogun, O. O. Oyerinde, and S. H. Mneney, "Performance Analysis of the OFDM-IDMA System with Carrier Frequency Offset in a Fast Fading Multipath Channel, in IEEE 3rd Wireless Vitae Conference, USA, June 24–27, 2013.

[15] L. Cimini. Analysis and Simulation of a Digital Mobile Channel Using Orthogonal Frequency Division Mutiplexing, IEEE Trans. Communication, vol. 33, no. 7, pp. 665–675, July 1985.

[16] A. Molisch. Orthogonal Frequency Division Multiplexing (OFDM), Wiley-IEEE Press eBook Chapters, second edition, pp. 417–43, 2011.

[17] Wireless LAN Medium Access Control (MAC) and Physical Layer (PHY) Specification: High-speed Physical Layer in the 5 GHz Band,

The Institute of Electrical and Electronics Engineers, Inc., IEEE Std. 802.1la–1999

[18] L. Liu,W. K. Leung, and Li Ping. Simple chip-by-chip multi-user detection for CDMAsystems," in Proc. IEEE VTC-Spring, Korea, pp. 2157–2161, Apr. 2003.

[19] K. Li, X. Wang, and L. Ping. Analysis and Optimization of Interleave-Division Multiple-Access Communication Systems, IEEE Trans. on Wireless Communications 2007, vol. 65, 1973.

[20] Q. Huang, K.-K. Ko, P. Wang, L. Ping, S. Chan. Interleave-division multiple-access based broadband wireless networks, Information theory workshop, pp. 502–506, 2006.

[21] L. Ping, L. Liu, K.Wu, and W. Leung. On interleave-division Multiple-Access, in IEEE International Conference on Communications, vol. 5, pp. 2869–2873, June 2004.

[22] L. Ping, Q. Guo and J. Tong. The OFDM-IDMA Approach to Wireless Communication System, IEEE Wireless Communication, pp.18-24, June 2007.

[23] L. Ping, L. Liu, and W. Leung. A simple approach to near-optimal multiuser detection: interleave-division multiple-access," in Proc. IEEE Wireless Comm. Networking (WCNC 2003), vol. 1, pp. 391–396, March 2003.

[24] B. Dongming, Y. Xinying. A new approach for carrier frequency offset estimation in OFDM communication system, IEEE communication Tech. Proc., ICCT 2003, vol.2, pp. 1922–1925, 2003.

[25] M. Morelli, C. Kuo, and M. Pun. Synchronization techniques for orthogonal frequency division multiple access (OFDMA): a tutorial review, Proc. IEEE, vol. 95, no. 7, pp. 1394–1427, July 2007.

[26] A. Al-Dweik and R. Hamila, "A highly efficient blind carrier frequency offset estimator for wireless OFDM systems," Proc. of the IEEE Int. Conf. on Consumer Electronics (ICCE '06), pp. 375–376, CA, USA, Jan. 2006.

[27] T. Peng, Y. Xiao, X. He and S. Li. Improved Detection of Uplink OFDM-IDMA Signals with Carrier Frequency Offsets, IEEE communication letter, vol. 165, pp. 646–649, May 2012.

[28] Y. Liu, X. Xiong, Z. Luo. Effect of Carrier Frequency Offsets on OFDM-IDMA Systems, 2012 2nd International Conference, pp. 209–302, 2012.

[29] L. Ping, L. Liu, K. Y. Wu, and W. K. Leung. Interleaved-Division Multiple-Access, IEEE Trans. Wireless Communication, vol.4, pp. 938–947, April 2006.

[30] T. Shan and T. Kailath. Adaptive algorithms with an automatic gain control feature," IEEE Transactions on circuits and systems, vol. 35, no. 1, pp. 122–127, January 1988.

[31] H. Modaghegh, R. H Khosravi, S. A Manesh and H. S Yazdi. A new modeling algorithm—Normalized Kernel Least Mean Square, IEEE International conference on Innovations in Information technology, IIT 2009, pp. 120–124, 2009.

[32] E. Alameda-Hernandez, D. Blanco, D. P Ruiz and M. C Carrion. The Averaged, Overdetermined, and Generalized LMS Algorithm, IEEE Transactions on signal processing, vol. 55, no 12, pp. 5593–5603, December 2007.

[33] Y. Rahmatallah, N. Bouaynaya and S. Mohan. Bit Error Rate Performance of Linear Companding Transforms for PAPR Reduction in OFDM Systems, in IEEE Global Communications Conference (GLOBECOM 2011), Houston, Texas, December 2011.

Biographies

Muyiwa Blessing Balogun received the B.Sc (Hons) in 2009 from the University of Ilorin, Nigeria. He is presently studying towards his M.Sc degree at the University of Kwazulu-Natal, Durban, South Africa. His research interests include frequency synchronization algorithms for multicarrier systems, multiple antenna systems and digital signal processing applications.

Olutayo Oyeyemi Oyerinde received the B.Sc. (Hons.) and the M.Sc. degrees in electrical and electronic engineering from Obafemi Awolowo University, Ile-Ife, Nigeria, in 2000 and 2004, respectively, and the Ph.D. degree in electronic engineering from the School of Engineering, University of KwaZulu-Natal (UKZN), Durban, South Africa, in 2010. He was a Postdoctoral Research Fellow with the School of Engineering, UKZN, under UKZN Postdoctoral Research Funding. He is currently a Telecommunications lecturer in the School of Electrical and Information Engineering, University of the Witwatersrand, South Africa. His research interests are in the area of wireless communications including multiple antenna systems, orthogonal frequency division multiplexing system and channels estimation, and signal processing techniques.

Stanley Henry Mneney received the B.Sc. (Hons.) Eng. degree from the University of Science and Technology, Kumasi, Ghana, in 1976 and the M.A.Sc. from the University of Toronto, Toronto, Ontario, Canada, in 1979. In a Nuffic funded project by the Netherlands government he embarked on a sandwich Ph.D programme between the Eindhoven University of Technology, Eindhoven, Netherlands and the University of Dares Salaam, Dares Salaam, Tanzania, the latter awarding the degree in 1988. He is at present a Professor of Telecommunication and Signal Processing and head of the Radio Access and Rural Telecommunication (RART) Centre in the School of Engineering, University of KwaZulu-Natal, Durban, South Africa. His research interests include theory and performance of telecommunication systems, low cost rural telecommunications services and networks, channel modelling and digital signal processing applications.

Efficient Fine Grained Access Control for RFID Inter-Enterprise System

Bayu Anggorojati, Neeli Rashmi Prasad, and Ramjee Prasad

Center for TeleInFrastruktur (CTIF) Aalborg University, Denmark
E-mail: {ba,np,prasad}@es.aau.dk

Received 24 August 2013; Accepted 12 November 2013;
Publication 23 January 2014

Abstract

Access control management is a very challenging task in an inter-enterprise RFID system due to huge amounts of information about things or objects that can be collected and accessed to and from the system. Furthermore, the information stored in the inter-enterprise RFID system contains sensitive and confidential data related to the activities of the organization involved around the RFID system. Hence, the efficiency and high-granularity are critical in the design of access control for such system. This paper presents a novel access control model which is efficient and fine grained for such a system. A detail definition and mechanism of the access control model are described in the paper. A system implementation is developed for the evaluation purpose. An important performance measure in big data processing is delay in processing time, thus the evaluation aims at measuring the access control processing time. The evaluation results show that the model is consistent, and is able to achieve less delay than the inter-enterprise RFID system without access control at a certain point.

Keywords: access control, policy, security, RFID, IoT.

1 Introduction

RFID technology allows the everyday things to be interconnected to the internet world, thus the key component towards the full deployment of the IoT vision [3]. From the system components and architecture perspectives, RFID

Journal of Cyber Security, Vol. 2 NO. 3 & 4, 221–242.
doi: 10.13052/jcsm2245-1439.232

system employed by any organization in its activity may consist of three sub-systems, namely RF, enterprise, and inter-enterprise sub-system [11]. Inter-enterprise sub-system in particular, is the most important component that enables the objects or things visibility and tracking throughout their life cycle, i.e. in supply chain industry, etc. When the information about thousands of things or objects is able to be gathered and accessed, consequently access control management of such information is a great challenge.

The technical specification, including the standard interfaces and data format, that enables the inter-enterprise information sharing of RFID events data is specified in the EPCIS specification [7] issued by EPC global. The EPCIS repository, i.e. software implementation of this specification, in particular aims at receiving application-agnostic RFID data, translate it into a corresponding business events (e.g. business process, business location, event time, etc), and then make the events available and accessible by upstream applications. Since the EPCIS repository potentially contains sensitive and confidential data of any individual or organization, the access to such information through its interfaces needs to be managed properly. In this regard, it is important to mention that the access control mechanism in the EPCIS specification is left open to each specific implementation. Additionally, efficiency, fine level of granularity, and trust are important keys to the access control design since highly dynamic and huge amount of events data is expected to be generated by potentially thousands of tagged objects which are of any interest for individuals or organizations that have even had any relationship before.

There are two main contributions of this paper. First, a dynamic and efficient access policy mechanism of an object or group of objects, based on the attributes and vocabularies of EPCIS is introduced. This access policy takes the profile of the accessing entity (i.e. individual or organization that requests to access RFID events information – this term will shortly be referred as *user* throughout the rest of this paper), and results in a suitable set of access rule of the corresponding entity to the object(s). This way, the access policy can be dynamically reuse for any user that even had no relationship before. Second, fine grained policy access enforcement method to handle large amounts of RFID events information, using the created rules and contextual information, is presented. For evaluation purpose, a system implementation of the proposed access control model is developed and tested.

The remainder of this paper is organized as follows: An overview of related works in access control is given in Section [2]. The problems, requirements, and realistic assumptions along with a real life example for designing an access control framework in RFID is described in Section [3]. The proposed access control framework, along with the definition of access policy of the object(s), mechanism to generate access rules for the accessing entity, and the access policy enforcement mechanism, is explained in Section [4]. The system implementation of the proposed access control model is presented in section [5]. The evaluation results and findings are discussed in Section [6]. Some qualitative discussion regarding important features of secure system and access control constituted in the proposed model as well as comparison with existing access control model are presented in Section [7]. Finally, the conclusion and future directions of this work are given in Section 8.

2 Related works

Study in various types of access control models have been quite well established within the computing and information technology field. Our particular interests are in incorporating the contextual information and dynamically create access rules based on a pre-defined set of policies. XACML [13] is an XML framework to describe access control policies for web based resources. The XACML specification incorporates some contextual information into access decisions, but it has no formal context-aware access control model. In addition, the access decision from the evaluated policies in XACML is only limited to four pre-defined categories, i.e. Permit, Deny, NotApplicable, or Indeterminate, which greatly reduces the granularity of access decision results.

Role Based Access Control RBAC [12] is an access control model that is widely used and further derived into different models, due to its suitability in almost any organization which consists of different roles with some levels of hierarchy. Temporal aspects of RBAC were addressed in TRBAC [14], which focuses on temporal availability and dependency of roles. GTRBAC [10] is an extension of TRBAC model that is capable in expressing a wide range of temporal constraints – in particular time periodicity as well as duration, and de/activating as well as enabling constraints – on roles. An XML specification of GTRBAC has been introduced in [6] and the extension of X-GTRBAC which incorporates trust in assigning roles to users has been presented in [5]. Although these models support context-awareness but the role based model, i.e. with user-to-role and role-to-permission mapping, does not fit with the requirement of RFID inter-enterprise system.

CCAAC [1], another type of access control model that supports contextual information and is based on capability. In addition, CCAAC provides a framework where a valid capability as a mean for an access request to be granted, is created for any user based set of access policies attached to an object or group of objects. Here, *object* refers to resource to be accessed by any user. The CCAAC offers efficient and dynamic way of managing access control through the evaluation of user's profile and contextual information via the corresponding access policies upon the capability request to certain object(s), which is important when dealing with huge numbers of objects and users simultaneously, e.g. in IoT or RFID system. Moreover, it also supports access delegation and revocation. However, the type of action and access decision result limits the level granularity, and the context-awareness is not formally modelled.

A fine grained access enforcement specially designed for the EPCIS events data through a rule-based policy language for Auto-ID events, called AAL, has been introduced in [9]. In addition, an efficient policy enforcement mechanism and implementation based on SQL query rewriting was presented. The main drawback of AAL as presented in [9] is that the access policy is manually assigned to users which is impractical in the real life situation. The dynamic generation, assignment, and revocation of access policies were not considered as well.

3 System requirements

The RFID events data in the EPCIS repository, which is known as EPCIS events, is categorized into four types of events namely *Object Event*, *Aggregation Event*, *Quantity Event*, and *Transaction Event* [7]. Each of them describes different type of event taking place in relation to the RFID tag, which is represented by EPC ID, in the business process within the company. In addition, two types of data, i.e. RFID application-agnostic and master/company data, are comprised in the EPCIS repository. Here, the master data, e.g. event Time, action, bizStep, etc, provides some necessary business context to interpret the EPCIS events [7].

A simplified example of an inter-enterprise RFID system deployment involving an EPCIS repository is depicted in Fig. 1. In this example, the EPCIS repository is owned by a company c_1, and is accessed by companies c_2 and c_3. The RFID events consisting of RFID and master data generated by a BEG module owned by c_1 is captured via capture interface and stored in the EPCIS repository as EPCIS events. The EPCIS events stored in EPCIS repository can then be accessed by other companies through the query interface. According to a comprehensive definition and description of the EPCIS specification [7], the

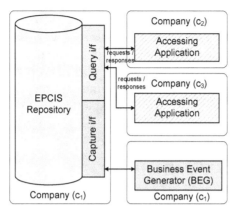

Figure 1   The interactions of EPCIS repository interfaces in an inter-enterprise RFID system

EPCIS events can be interpreted and give a valuable information regarding the business activities of company c_1 by business context information and RFID data. For example, the interpreted EPCIS event can give a figure about production volume, sales activity of certain products, inventory status, etc. Obviously, these are very sensitive information that a company want to reveal as minimum as possible to other parties by managing access to EPCIS event data in high level of granularity. In addition, the the RFID tags' owner may want to restrict the access to particular information related to the activity of the tags which might reveal another type of sensitive information apart from the business related information.

Based on the identified problems, the following requirements for access control in an inter-enterprise RFID system should be foreseen:

- **Context-awareness:** The access control system should be design to support rich business context information that are contained in the EPCIS events.
- **Dynamic rule:** Providing a fine grained access control in a highly dynamic EPCIS events data that is generated continuously, it is almost impossible to assign a static access rights for particular users to certain part of data or attributes. Therefore, dynamic access rule should be generated based on the specified rules in the access policy and the requested set of information. In addition, the policy should support flexible inclusion of new events.
- **Dynamic access assignment:** There are certainly various types of users that are trying to gain access to the EPCIS events data which are probably

not known before by the EPCIS repository's system administrator, i.e. the responsible person to create the access policy. Hence, a dynamic mechanism to assign access rights to users is also required.

- **Object based policy:** The access policy based on particular RFID tag IDs, e.g. on the Object Class or serial numbers level, is also required to restrict the access to information related to tag's activities by the tag's owner.

4 Proposed Access Control Model

4.1 Assumptions

Based on the identified problems and how the inter-enterprise RFID system operates, the following assumptions are made as a baseline to design an efficient, dynamic, fine grained access control for inter-enterprise RFID system:

- The authentication phase has been carried out before the access control process takes place.
- The EPCIS events stored in the EPCIS repository are only events generated by the EPCIS repository's owner, e.g. a company.
- The set of contextual information, i.e. attributes of EPCIS events, such as bizStep, action, disposition, readPoint, epcList, etc, and their values are known to the EPCIS repository's owner. The values of some attributes are always fixed, e.g. $action = \{'ADD','OBSERVE','DELETE'\}$, while the values of attributes like *epcList* are dynamics but the company has a full knowledge of all EPC IDs involved in their business transactions.
- The EPCIS owner may or may not know the users that are accessing its EPCIS events through the query interface. But the user's profile, such as company name, location, business area, etc, can be obtained through a trusted means, e.g. via trusted third party organization.
- For each query of EPCIS events, the user may optionally specify particular Event Type(s) and some contexts or event attributes, e.g. event Time, action, bizStep, etc, as stated in the EPCIS specification [7]. It is important to note that the default query operation according to the EPCIS specification is that all the available information will be returned unless specified otherwise in the query.
- The RFID tag's owners have the knowledge of their tags' IDs in order to create access policies for an individual tag or a set of tags.

These assumptions lead us to propose an access control model that will be explained in the following subsections.

4.2 Definitions

4.2.1 Elements, Attributes, and Values

First of all, the data structure in our proposed access control framework is based upon *Element* → *Attribute* → *Value* ternary relationship which maps each element to its attributes and their values. *Element* is a set of elements which could be user profile, P, or EPCIS event, E. Each of the *element* consists of several attributes and each of the attribute can have a set of possible values.

4.2.2 EPCIS Event

Let E_i be the ith EPCIS event within the set of EPCIS events element. As mentioned previously, according to [?] E_i could be an *ObjectEvent*, *AggregationEvent*, *QuantityEvent*, or *TransactionEvent*. Each of E_i consists of a set of attributes as defined in details in [?]. Let us call the jth attribute of an E_i as AE_j. Here, the relationship between E_i and AE_j can be expressed in the following notation: $E_i = \{AE_1, \cdots, AE_n\}$.

The value of each AE_j, let us call it as VE_{jk}, can either be a single value or a list of values, e.g. in the case of the *epcList*. Moreover, the value of some attributes may be among some pre-defined values. In any case, the relationship between AE_j and VE_k can be generally expressed as $AE_j = \{VE_{j1}, \cdots, VE_{jn}\}$.

4.2.3 User profile

In addition to our previous assumption, the administrator can define a set of user profile based on several attributes and values. Now, let P_i be the ith user profile among a set of user profiles defined by the administrator. Similar to E_i, each P_i consists of several user profile attributes, e.g. AP_j, and each AP_j may consists of a set of values, i.e. VAP_{jk}, defined by the administrator.

$$AP_j = \{VAP_{j1}, \cdots, VAP_{jn}\} \tag{1}$$

4.2.4 Objects

The *Object* field can be expressed as a single tag ID or as a pattern which can apply to a set of tags, according to the TDS specification [8]. In mathematical form, the *Object* O can simply be expressed as a set of object values: $O = \{VO_i, \cdots, VO_n\}$

4.2.5 Condition

Condition is the key component to provide dynamic rule and fine-grained access control. It is important to mention that the *condition* in this proposed model is not a requirement for granting an access or not. Rather, it is defined as a set of constraint applied to the contextual information. In our case, the contextual information is specific to the business contexts or the EPCIS event attributes (AE_j).

The way the condition is expressed in the proposed access control policy follows the white-listing principle, meaning that the access to information that is not explicitly mentioned in the condition is not allowed, which is the opposite of the nature of the EPCIS query specification as explained earlier. This principle is valid in general, and especially in our case where the EPCIS event attributes are defined in great details.

With this principle in mind, a condition can be expressed in general as subset of AE_j, i.e. $AE_j^s \subseteq AE_j$, with respect to a set of all possible values of a particular AE_j. This implies that a condition can set whether to allow all, partly, or none of the values to be accessible. For example, let say an EPCIS event attribute consists some pre-defined values, e.g. $AE_j = \{a, b, c\}$. Some possible condition for such an attribute AE_j are $AE_j^s = AE_j = \{a, b, c\}$, $AE_j^s = \{b, c\}$, or $AE_j^s = \varnothing$, which means that the values of AE_j is shown fully, partially (e.g. only $\{b, c\}$), and none. This way of describing the condition is also to fulfil the purpose of maintaining the consistency upon the existence of multiple rules.

On a practical stand point in writing a condition, there are two possible ways to specify it: first, by explicitly stating what set of values, i.e. VE_{jk} are accessible; second, by describing values in certain ranges using a comparison operator. The first way is more static and suitable for attributes that have some pre-defined values, e.g. action, business location, disposition, etc, whereas the later one is more dynamic and capable of specifying a condition for event that has not happened yet, such as the *event Time*. In such a case, general statement of the condition of the rule i with respect to AE_j^s is as follows:

$$C_i = AE_{i1}^s \wedge \cdots \wedge AE_{ij}^s \wedge \cdots \wedge AE_{in}^s \tag{2}$$

4.2.6 Rule

A *rule* consists of a set of *conditions*, C and an EPCIS event type, ET. General representation of a rule using a **if**-**then** relationship is: $ET \Rightarrow C$. A set of rules together with a user profile P or a set of objects O forms a policy that will be explained in the next Subsection (4.2.7).

4.2.7 Policy

In the proposed model, there are two types of *policies*, namely user based and object based policies. A *user based policy* consists of a set of *profiles* P and a set of rules, while an *subject based policy* consists of a set of *objects* O and a set of rules. A set of different rules that are applicable for a *request*, i.e. matched to the *request*'s P or O, and having the same ET value are combined into a new applicable rule using a *disjunction* operation after resolving all the conflicting policies. In addition, multiple policies may also be applicable to an access *request*, thus a policy combining mechanism should take place in order to create a final applicable rule for an access *request*. Finally, the final applicable rule would then be evaluated with the *conditions* presented within the access *request* in order to get a final access decision. Please note that the access decision is not in a form of static *permit* or *deny*, rather it is in a set of access constraints or conditions that explicitly authorize which data can be accessed by the user. The whole mechanisms of the rule and policy combining, and then evaluate them with the access *request* are explained in great details in the Subsection 4.3.

4.2.8 Request

An access *request* must consists of a *profile* P, and optionally a set of EPCIS event types ET and a set of *conditions* C. Both ET and C are optional fields in a *request* due to the nature of the EPCIS query specification. For a *request* to be evaluated with a certain policy, the *request*'s profile P_{REQ} should match with the *policy*'s profile P_{pol}. If a single or set of ET(s) are specified in the *request*, then only rules that matched with the specified ETs are applicable, otherwise all rules in the policy are applicable for the *request*.

4.3 Access Rules Evaluation

Access rules evaluation is at the heart of the whole model in order to provide a fine-grained access control. The results of this evaluation is a final access decision in a form of a set of explicit access constraints.

There are two important steps in the access rules evaluation. First, all applicable rules are combined by using union operation. Second, the combined rules' condition is then evaluated against the condition presented by access request using an intersection operation. The general expression of this evaluation is depicted as follows:

$$C^{REQ} \cap (C_1 \cup \cdots \cup C_p) \tag{3}$$

Algorithm 1 access rules evaluation procedures

Procedure RulesEvaluation(C^{REQ}, SC)
    for all AE^s_{ij} in C_i in SC do
        for all C_i in SC do
            $AE^{s'}_j = UnionAll(AE^s_{xj})$
        end for
        Combine Rules = $Conjunction(AE^{s'}_j)$
    end for
    $Condition = ""$
    for all $AE^{s'}_j$ in *Combine Rules* do
        if $Condition \neq ""$ then
            $Condition \mathrel{+}= \wedge$
        end if
        for all AE^r_k in C^{REQ} do
            if $j == k$
                $Condition \mathrel{+}= AE^{s'}_j \cap AE^r_k$
            end if
        end for
        if not Found($k == j$) then
            $Condition \mathrel{+}= AE^s_j$
        end if
    end for
end procedure

---

where:

- p: number of matched access rules.
- $C_i = (AE^s_{ij} \wedge \cdots \wedge AE^s_{in})$: a set of conditions specified in the ith matched access rule.
- $C^{REQ} = \{AE^r_k, \cdots, AE^r_m\}$: a set of EPCIS attributes conditions specified in user request.

The access rules evaluation as expressed in (3) is obtained through some steps illustrated in the pseudo-code 1.

The procedure starts by combining all conflict-free conditions of the applicable rules using union set operation. Once a combined rule is obtained, the access rule evaluation is started by grouping the similar event type attribute, i.e. AE pair, from between the C^{REQ} and the $CombineRules$, and then a set intersection operation is applied, i.e. $AE^{s'}_j \wedge AE^r_k$, where $j = k$. In case of the pair of $AE^{s'}_j$ is not specified in the C^{REQ}, then only the condition as specified in AE^s_j will be applied. At last, the final expression of access rule evaluation as depicted in (3) is obtained.

5 System Implementation

Two important components in this system implementation are the Policy repository as well as evaluation, and Access Policy enforcement components. In addition, the overall implementation is designed to be modular and complies with the EPCIS query interface specification. For this purpose, the proposed access control is implemented through web service interface and be part of the Aspire RFID middleware [2] module. The Aspire RFID middleware itself is one of the open source implementation of RFID middleware based on EPCglobal specifications.

5.1 System Architecture

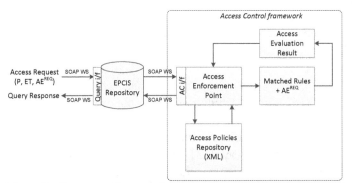

Figure 2 System architecture of the implemented access control model

The system architecture of the implemented access control model is depicted in Fig. 2. The Access Control Framework is communicating with the EPCIS repository through web service interface. Once the incoming query or access request is received, the access policy enforcement point will query the Access Policies Repository to find the matching policies with the request according to the procedure explained in Section 4 earlier. Once a set of matching policies or rules is found, it will be evaluated against the access request. Finally, the result of the access control evaluation will be sent back to EPCIS repository through the Access Enforcement Point via the web interface.

The access policy enforcement is an important component in a policy based access control implementation after an access decision has been taken. In our implementation, the access control policy enforcement is implemented in a form of a modified *Query Params*, i.e. the query parameters defined in the EPCIS specification. The advantage of this method is that the implementation complies fully with the EPCIS query interface specification while stay modular without a necessity to greatly modify the EPCIS server implementation. Moreover, it does not depend of the actual implementation of the database technology for the EPCIS repository. However, the main drawback is that this method does not fully leverage the proposed access control model due to the nature of *Query Params* description in the EPCIS specification where the parameter that is not explicitly mentioned in the *Query Params* will be included in the query results. Nevertheless, it still allows us to perform an evaluation over most of the important functionalities of the proposed access control model. Additionally, if the access decision return an empty result, i.e. no matching policy is found in the repository, the query results of the EPCIS query interface implementation will return a *Security Execption*.

The policy repository and evaluation component implementation uses XML and JAXB technologies. The proposed access control policy specification as explained in Section 4 is translated into a set of XML specification and is practically stored as an XML file. Thanks to the JAXB technology, the access control policies can be created as an XML file and the stored access control policies can be translated into Java Objects to be further evaluated.

The usage of XML for describing policies in the proposed access control model is motivated due to the fact that the XML has been well accepted and widely used in the heterogeneous IT enterprise environment across different platforms. Among others, XACML [13] is the most famous XML based access control system. Particularly in our system implementation, the XML is used it model the data structure of the proposed access policy elements as described in Section 4.

6 Evaluation Results and Discussion

A series of experiments or evaluation have been carried out based on our implementation in order to measure the performance of the proposed access control model. The evaluation encompasses two main purposes. First, it is meant to validate the functionality of the proposed access control model. Second, it aims at measuring the performance in terms of delay time. Delay is an important parameter to be evaluated due to the fact that time is a critical performance metric in dealing with big data processing.

6.1 Evaluation Procedures

For the testing purpose, one policy for a particular user profile is prepared. The policy consists of different rules with several conditions for different EPCIS event types. It is important to note that since generating EPCIS events already requires some complex procedures, we focuses on generating only one type of business event in the test scenario. Hence, the generated EPCIS events are always having fixed event attributes values ($V E_i j$), except for the *event Time*, *record Time* and EPC ID. As a result, the final access control enforcement mainly depends on conditions related to *event Time* and EPC ID parameters or attributes in the practical experiments.

Regarding the first objective of this evaluation, it is shown that the access control works as it should be. The query results only returns the EPCIS event data that fulfil the conditions set in the policy and the final access decision. On the other hand, a *Security Exception* will be returned if there is no matching policy found in the access policy repository.

Concerning the second objective of the evaluation, the delay performance of the proposed access control model is compared against the EPCIS system without any access control applied. For this purpose, we have tested our proposed access control model on a system with Intel Core i7 2.80 GHz CPU, 8 GB RAM, running Windows 7, and Java 6.31. In addition, an Apache Tomcat 6.35 server is used to deploy all the Aspire RFID Middleware modules as well as our access control module, and a MySQL 5.2 Server is used as the EPCIS events data repository. Although the measurement results strongly dependant on the implementation, nevertheless we can expect to draw some qualitative conclusion out of the results and further improve the system implementation if necessary.

Two measurement scenarios are carried out to observe the delay performance and behaviour of the implemented proposed access control model. The delay is defined as the time consumed between the query request being sent and received by the EPCIS query client, i.e. $T_{received} - T_{sent}$. For each scenario, the same measurement point is done repeatedly for 700 times. For the purpose of explanation in the rest of this paper, we will refer the case when the access control is used as the *first case* and the *second case* refers to the case when access control is not used.

6.2 The Impact of Varying the Number of Read Tags

The first scenario aims to observe the delay time behaviour when the number of read tags in each event stored in the EPCIS repository is varied. In this scenario, the number of EPCIS events are fixed to three events at the EPCIS repository. It should be noted that the query results of the *first case* always returns one EPCIS events, while the *second case* always returns all three events since no access control is applied.

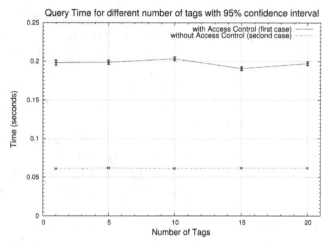

Figure 3   Query delay time for different number of tags with 95% confidence interval

There are two important findings that can be derived based on the results of this scenario which is shown in Fig. 3. First, the average delay of the *first case* is three times higher than that of the *second case*. This finding is quite expected because it involves extensive XML processing, e.g. creating and parsing of SOAP message and parsing the access policy, which is time consuming. The delay performance could be improved in general through a more tightly coupled implementation with the expense of sacrificing the flexibility and modularity of such an open system. Second, the EPCIS query delay time for both cases is relatively constant regardless of the number of tags read in each EPCIS event. Although some small variation is shown in the *first case*, the confidence interval margin is rather high and it could be due to the inconsistency of web service invocation of access control service. Thus it is quite safe to neglect the small variation. Nevertheless, the second finding is quite interesting because the amount of data queried from the database would contribute to the query time delay.

6.3 The Impact of the Number of the EPCIS Events Data in the Repository

Based on the second finding in the first scenario, we would like to check the impact of varying the number of EPCIS events in the repository to the query time delay. Knowing that the number of tags does not change the delay behaviour, the number of tag included in the EPCIS events stored within the repository is fixed to only one tag in this scenario. Similar to the first scenario, the *first case* always returns one EPCIS events while the *second case* always returns all the EPCIS events available in the repository.

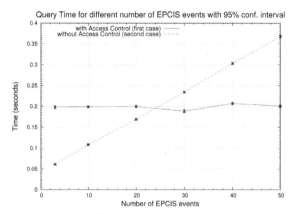

Figure 4   Query delay time for different number EPCIS events in the repository with 95% confidence interval

The results of the second scenario depicted in Fig. 4 shows that the number of EPCIS events data queried from the repository does impact the query time delay linearly. Obviously, the query delay time of the *first case* is relatively constant since it always return one EPCIS event as a result of the access control mechanism. Based on this finding, it can be concluded that both cases would achieve the same query delay time when the ratio of the EPCIS events data returned between the *first* and *second* case is around 1 : 25. Consequently, the delay gap between both cases as shown in Fig. 3 would be bigger if the result of access control evaluation would return more than one EPCIS events data. This ratio could be improved through a more efficient implementation.

7 Discussions

Earlier in this paper, we have mentioned several requirements of access control model in an inter-enterprise RFID system, and how the proposed model fulfilled those requirements. In this section, we will present several important features of a secure system – in particular an access control for big data system – that are constituted in the proposed access control model. Furthermore, some qualitative comparisons with other access control models will also be given.

- **High-granularity:** The proposed access control model provides high-granularity since it does not allows the access only based on *permit* or *deny* action, but based on some specific types of data or context information. The level of granularity can be defined in the access policy by the system or security administrator of the system.
- **High privacy:** The high-granularity of access control policy provides high level of privacy to the business activities related information stored in the EPCIS repository.
- **Trust:** The access policy is applied to each user automatically based on some pre-defined user profiles, and a user is assigned to a particular user-profile based on trust relationship and contextual information.
- **Flexibility:** The proposed access control policy model offers high flexibility which is favourable in a big data system. It allows relative time definition instead of fixed time, EPC ID definition based on the EPC pattern, and automatic user assignment to user-profile through trust information.
- **Inter-operability:** The proposed access control model complies with the EPCIS Specification which is an open standard. The proposed system is also implemented as a web service which is highly inter-operable, i.e. independent of any specific programming language or server implementation.

In comparison with other existing access control, the proposed access control model is better as compared to them in the following aspects:

- *Trust-based X-GTRBAC[5]:* Our proposed access control model allows a way to incorporate trust in assigning user-profile to users which is quite similar to the approach presented in [15]. However, [15] does not fit the requirement of providing high-granularity access to data, particularly in an inter-enterprise RFID system.
- *AAL [9]:* A quite similar approach of a fine grained access enforcement specially designed for the EPCIS events data that is proposed in our access control model, was also introduced in [9]. However, automatic way of assigning some access policy to a user was not considered in [9], which gives a lot of burden to the system administrator, thus unrealistic for the real system. Moreover, its SQL query rewriting method for the access enforcement, does not inter-operable with the EPCIS implementation that is not based on the RDBMS.

8 Conclusion

Access control management in an inter-enterprise RFID system is a great challenge, since such system allows the tracking and monitoring of large numbers of things, i.e. a path towards realizing the IoT vision. The most challenging access control problems in such a system are in providing high-granularity access of RFID events data, known as EPCIS events, with flexibility and efficient access policy management over a very dynamic and huge amounts of EPCIS events data. A novel access control model to address these problems has been proposed in this paper along with complete definition of access control model and evaluation through the system implementation of the model. The findings in the evaluation show that the proposed access control model is consistent and could achieve less query time delay than the inter-enterprise RFID system without access control if the ratio of the returned EPCIS events data is less than $1 : 25$.

The proposed access control model relies on the trusted third parties entity to obtain the user profile, but the mechanism on dealing with such trust management has not been addressed yet. Incorporating trust into the access control model, i.e. through PKI CA, is one possibility of the future work. Another direction targeted in the near future is to improve the current system level implementation, especially in addressing the limitation described in Section 5.

Reference

[1] Anggorojati, P. N. Mahalle, N. R. Prasad, and R. Prasad. Secure access control and authority delegation based on capability and context awareness for federated iot. In Fabrice Theoleyre and Ai-Chun Pang, editors, *Internet of Things and M2M Communications*. River Publisher, 2013.

[2] ASPIRE. http://wiki.aspire.ow2.org.

[3] L. Atzori, A. Iera, and G. Morabito. The internet of things: A survey. *Computer Networks*, 54(15):2787–2805, 2010.

[4] E. Bertino, P. A. Bonatti, and E. Ferrari. Trbac: A temporal role-based access control model. *ACM Trans. Inf. Syst. Secur.*, 4(3):191–233, August 2001.

[5] R. Bhatti, E. Bertino, and A. Ghafoor. A trust-based context-aware access control model for web-services. *In Web Services, 2004. Proceedings. IEEE International Conference on*, pages 184–191, july 2004.

[6] R. Bhatti, A. Ghafoor, E. Bertino, and J. B. D. Joshi. X-gtrbac: an xml-based policy specification framework and architecture for enterprise-wide access control. *ACM Trans. Inf. Syst. Secur.,* 8(2): 187–227, May 2005.

[7] EPCglobal. Epc information services (epcis) version 1.0.1 specification. September 2007.

[8] EPCglobal. Gs1 epc tag data standard 1.6 - ratified standard. September 2011.

[9] E. Grummt and M. Muller. Fine-grained access control for epc information services. In *Proceedings of the 1st International Conference on The Internet of Things,* IOT'08, pages 35–49, Berlin, Heidelberg, 2008. Springer-Verlag.

[10] J. B. D. Joshi, E. Bertino, U. Latif, and A. Ghafoor. A generalized temporal role-based access control model. Knowledge and Data Engineering, *IEEE Transactions on,* 17(1):4–23, jan. 2005.

[11] T. Karygiannis, B. Eydt, G. Barber, Lynn Bunn, and T. Phillips. Guidelines for securing radio frequency identification (rfid) systems - recommendations of the national institute of standards and technology. *NIST Special Publication,* April 2007.

[12] R. S. Sandhu, E. J. Coyne, H. L. Feinstein, and C. E. Youman. Role-based access control models. *Computer,* 29(2):38–47, feb 1996.

[13] XACML. https://www.oasis-open.org/standards#xacmlv2.0

Acronyms

6LoWPAN IPv6 over Low-power Wireless Personal Area Network
AAL Auto-ID Authorization Language
ACL Access Control List
ACS Access Control Servers
AVISPA Automated Validation of Internet Security Protocols and Applications
BEG Business Event Generator
CASM Context Aware Security Manager
CA Certificate Authority
CA Certification Authority
CCAAC Capability-based Context Aware Access Control
CWAC Context aWare Access Control
DoS Denial of Service
EPCIS Electronic Product Code Information Services
EPC Electronic Product Code
FM Federation Manager
GTRBAC General Temporal RBAC
GW Gateway
ICAP Identity based Capability
IdP Identity Provider
IoT-DS IoT Directory Service
IoT-FM IoT Federation Manager
IoT Internet of Things
ITU International Telecommunication Union
JAXB Java Architecture for XML Binding
MAC Message Authentication Code
MAGNET My Adaptive Global NETwork
NFC Near Field Communication
OS Object Service
PE Policy Engine
PKI Public Key Infrastructure
PN-F Personal Network Federation
PNDS Personal Network Directory Service
PNDS PN Directory Service
PN Personal Network
PTD Personal Trusted Device
RBAC Role Based Access Control
RDBMS Relational Data Base Management System
RFID Radio Frequency Identification

RF   Radio Frequency
SDP   Security Decision Point
SecaaS   Security as a Service
SOAP   Simple Object Access Protocol
SQL   Structured Query Language
TDS   Tag Data Standard
TRBAC   Temporal RBAC
VID   Virtual Identity
WPAN   Wireless Personal Area Network
WSN   Wireless Sensor Network
XACML   Extensible Access Control Markup Language
XML   eXtensible Markup Language

Biographies

Bayu Anggorojati received his B.E. degree in Electrical Engineering in 2005 from Institut Teknologi Bandung, Bandung, Indonesia. He received his MSc in Mobile Communication in 2007 from Aalborg University, Aalborg, Denmark. He joined Wireless Security and Sensor Network group within Network and Security Section (now CTIF Section) in the Electronic System of Aalborg University as a research assistant from 2007 till now. He has been involved in a number of EU-funded R&D projects, including FP7 CP Betaas for M2M & Cloud, FP7 CIP-PSP LIFE 2.0, FP7 IP ISISEMDICt for Demetia, and FP7 IP ASPIRE RFID and Middleware. He is currently pursuing his PhD degree at CTIF Section in Electronic System Department of Aalborg University, Denmark. His research interests include Radio Resource Management in OFDMA based system; Access Control, Authentication, and Key Management in the IoT/M2M and Cloud system.

Neeli Prasad is leading a global team of 20+ researchers across multiple technical areas and projects in Japan, India, throughout Europe and USA. She has a Master of Science degree from Delft University, Netherlands and a PhD degree in electrical and electronic engineering from University of Rome Tor Vergata, Italy. She has been involved in projects totaling more than $120 million – many of which she has been the principal investigator. Her notable accomplishments include enhancing the technology of multinational players including Cisco, HUAWEI, NIKSUN, Nokia-Siemens and NICT as well as defining the reference framework for Future Internet Assembly and being one of the early key contributors to Internet of Things. She is also an advisor to the European Commission and expert member of governmental working groups and cross-continental forums. Previously, she has served as chief architect on large-scale projects from both the network operator and vendor side looking across the entire product and solution portfolio covering wireless, mobility, security, Internet of Things, Machine-to-Machine, eHealth, smart cities and cloud technologies. She has more than 250 publications and published two of the first books on WLAN. She is an IEEE senior member and an IEEE Communications Society Distinguished Lecturer.

Ramjee Prasad is currently the Director of the Center for TeleInfrastruktur (CTIF) at Aalborg University, Denmark and Professor, Wireless Information Multimedia Communication Chair. Ramjee Prasad is the Founding Chairman of the Global ICT Standardisation Forum for India (GISFI: www.gisfi.org) established in 2009. GISFI has the purpose of increasing of the collaboration between European, Indian, Japanese, North-American and other worldwide standardization activities in the area of Information and Communication Technology (ICT) and related application areas. He was the Founding Chairman of the HERMES Partnership - a network of leading independent European research centres established in 1997, of which he is now the Honorary Chair. He is the founding editor-in-chief of the Springer International Journal on Wireless Personal Communications. He is a member of the editorial board of other renowned international journals including those of River Publishers. Ramjee Prasad is a member of the Steering, Advisory, and Technical Program committees of many renowned annual international conferences including Wireless Personal Multimedia Communications Symposium (WPMC) and Wireless VITAE. He is a Fellow of the Institute of Electrical and Electronic Engineers (IEEE), USA, the Institution of Electronics and Telecommunications Engineers (IETE), India, the Institution of Engineering and Technology (IET), UK, and a member of the Netherlands Electronics and Radio Society (NERG), and the Danish Engineering Society (IDA). He is also a Knight ("Ridder") of the Order of Dannebrog (2010), a distinguishment awarded by the Queen of Denmark.

Dynamic AES – Extending the Lifetime?

Henrik Tange[1] and Birger Andersen[2]

[1]Aalborg University, Frederik Bajers Vej 7, DK-9220 Aalborg, Denmark,
het@es.aau.dk
[2]Center for Wireless Systems and Applications / CTIF-Copenhagen, Technical
University of Denmark, DTU Ballerup Campus, DK-2750 Ballerup, Denmark,
birad@dtu.dk

Received 15 October 2013; Accepted 6 December 2013;
Publication 23 January 2014

Abstract

AES (Advanced Encryption Standard) is a worldwide used standard for symmetric encryption and decryption. AES is for instance used in LTE (Long-Term Evolution) and in Wi-Fi. AES is based on operations of permutations and substitutions. Furthermore, AES is using a key scheduling algorithm. It has been proven that AES is vulnerable to side-channel attacks, related sub-key attacks and biclicque attacks. This paper introduces a new dynamic version of AES where the main flow is depending on the TNAF (τ-adic Non-Adjacent Form) value. This new approach can prevent side-channel attacks, related sub-key attacks and biclique attacks.

Keywords: AES, side-channel attacks, attack countermeasures, TNAF, ECC, related sub-key attacks, biclique attacks.

1 Introduction

The Rijndael algorithm was in 2001 selected by NIST to be the successor to DES (Data Encryption Standard) as AES [1]. The AES algorithm is based on finite mathematics, but there exists no mathematical proof. The AES was until recently considered secure.

The AES algorithm uses a fixed block size of 128 bits and different key sizes of 128, 192 or 256 bits [1, p.14]. Internally AES is using a state

Journal of Cyber Security, Vol. 2 No. 3 & 4 , 243–264.
doi: 10.13052/jcsm2245-1439.233

array on 4 x 4 bytes. AES is a Non-Feistel network [2, p.8]. A Non-Feistel network is using the two different operations for encryption and decryption. In AES a reverse algorithm for decryption is used. The four encryption operations are: AddRoundKey, SubBytes, ShiftRows and MixColumns. The four reverse operations are: AddRoundKey, InvSubBytes, InvShiftRows and InvMixColumns. The use of four encryption operations follows a well-known described scheme in the main algorithm consisting of rounds: In the initial round AddRoundKey is performed. In the following rounds (let's call them center-rounds) SubBytes, ShiftRows, MixColumns and AddRoundKey are performed. In the last round only SubBytes, ShiftRows and AddRoundKey are performed. In the decryption algorithm the order is: InvMixColumn, AddRoundKey, InvSubBytes and InvShiftRow. In both the encryption algorithm and the decryption algorithm the state array is containing the result from each operation. If the key size is 128 bits 10 center-rounds are executed; if the key size is 192 bits, the number of center-rounds is 12 and finally, if the key size is 256 bits, the number of center-rounds is 14.

The implementation of AES is fairly simple. It only requires table lookups (S-Box, an inverse S-Box and a Galois field multiplication array), shift operations and XOR operations. The AES algorithm can thereby be considered as a mix of substitutions and permutations.

A side-channel attack can be defined as an attack exploiting emitted information which is not intended to be used in the main operation [3, p.1].

A related sub-key attack can be performed by for instance a boomerang attack. A boomerang distinguisher can be found by searching for a local collision in the cipher [6, p.3].

A biclique attack is considered 3 to 5 times faster than a brute force attack [7, p.3]. A biclique attack can be based on the meet-in-the-middle principle. The attacker chooses an internal variable in the transform of data as a function of a plaintext and a key identical for all keys in a row and as a function of a ciphertext and a key identical for all keys in a column.

In the following subsections, A-C, we are further defining and discussing these three types of attacks.

In section II we are discussing related work and in section III we are presenting our contribution which is the extension of AES into a dynamic AES by introducing dependency on the TNAF value. This way we are addressing the three types of attacks. In section IV we shortly describe an implementation of dynamic AES, whereas section V presents tests and results. Finally, we analyze results in section VI and conclude in section VII.

1.1 Side-Channel Attacks on AES

The side-channel attack investigates the state array given a plaintext or a ciphertext, also called a known-plaintext attack and known-ciphertext attack, and a key. Another variant is to extract the key without knowledge about the plaintext or ciphertext. A practical attack can be done by having access to the data bus or specialized hardware making it possible to read the cache. In 2005 it succeeded for Osvik, Shamir and Tromer (OST) to perform a side-channel attack using the CPU memory cache [4]. This attack is possible since there is memory access to all tables in AES including the state array.

A type of a side-channel attack is a timing attack. This kind of attack has been shown by Joseph Bonneau and Ilya Mironov [5]. They show a model for attacking AES using timing effects of cache collisions. Cache is a near memory area between the CPU and the main memory. A cache collision is defined as when two separate lookups l_i, l_j where $l_i = l_j$. If $l_i \neq l_j$ it will result in a cache miss [5, p.206]. The assumption is therefore that the average time when $l_i \neq l_j$ is higher than the case when $l_i = l_j$ because it will cost a second cache lookup [5, p.205].

In a first round attack the attacker analyzes table lookups where the indices $x_i^0 = p_i \oplus k_i$ where p is a plaintext byte and k_i is a key byte. The bytes $\{x_0^0, x_4^0, x_8^0, x_{12}^0\}$ is a family of four bytes and are used as an index into table T_0. Three other families of bytes share the tables T_1, T_2, T_3 in round one. The attacker will have four sets of equations for each table, where each table will consist of a redundant set of six equations. However, there is no way to gain the exact key information. The attacker has to guess a value for one complete byte in each table family. The attack has succeeded with an average of $2^{14.6}$ timing samples [5, p.207].

A final round attack is using the algorithm fact that the MixColumn function is omitted in the final round. Thereby the equation is creating the ciphertext C by a simple lookup in AES S-box. The non-linearity in the AES S-box is the reason that this attack will succeed [5, p.208].

The main goal in a final round attack is to construct a guess at the final 16 bytes of the expanded key in the presence of noise [5, p.208]. Given the final key bytes it is possible to reverse the key expansion algorithm to find the original private key k.

1.2 Related Sub-key Attacks on AES

A related key attack is for instance performed by Alex Biryukow and Dmitry Khovratowich [6]. In this attack type a boomerang switching technique is used.

The attacker uses a pair of plaintexts (X_0, X_1) with a known difference α and encrypts both. Then the two ciphertexts (Z_0, Z_1) are both added a difference δ. This results in two new plaintexts (X_2, X_3). The four plaintexts form a quartet if $X_2 \oplus X_3 = \alpha$. Now the differences δ in the two pairs (Z_0, Z_2) and (Z_1, Z_3) are converted to the difference γ in the pairs (Y_0, Y_2) and (Y_1, Y_3) with probability q^2. If $Y_0 \oplus Y_1 = \beta$ the intermediate texts also form a quartet. Finally the pair (Y_2, Y_3) is decrypted with difference α with the probability p. A pair will result in a quartet with probability $p^2 q^2$. If $p^2 q^2 > 2^{-n}$, where n is the number of bits, a boomerang distinguisher exists.

The attack is possible since the key scheduling in AES is close to linear and therefore the subkeys can be viewed as a codeword of a linear code [6, p.3].

1.3 Biclique Attacks on AES

A biclique attack is based on bipartite graph known from graph theory. This attack was performed by Andrey Bogdanov, Dmitry Khovratovich, and Christian Rechberger at Microsoft Research [7]. A biclique is formed by the number of rounds and dimension. There exist two paradigms for key recovery using biclique. The first is called long biclique. A long biclique can for instance be constructed as a local collision. The second paradigm is called independent biclique. It is based on a high dimension for smaller $b < r$-m number of rounds, where m is the meet-in-the-middle attack out of r rounds. The smaller number of rounds makes it easier and with the use of simpler tools to construct a biclique [7, p.3].

The biclique attack can be applied to all versions of AES [7, p.3]. This type of attack can be up to a factor 5 faster for a key recovery of a round-reduced AES variant compared to a brute force attack.

The simple biclique attack will only require one plaintext-ciphertext data pair. In the meet-in-the-middle attack the attacker chooses a key space partition and places it into groups of keys with cardinality 2^{2d}. The key is now indexed as an element into a 2^d x 2^d matrix: $K[I, j]$. From the data transformation of the plaintext (P) a variable V can now be chosen such that:

$$P \xrightarrow[f_1]{K[i, -]} V \tag{1}$$

This is a function of the plaintext and a key identical for all keys in a row.

As a function of the ciphertext (C) and a key, we get this:

$$V \leftarrow \frac{K[-,j]}{f_2} C \tag{2}$$

This function is identical for the ciphertext C and a key for all keys in a column. The parts f_1 and f_2 correspond to the same parts of the ciphertext.

Now having the pair (P, C) the attacker can now compute 2^d possible values of $V \leftarrow$ and $\rightarrow V$ from the plaintext part and the ciphertext part. The meet-in-the-middle attack is more effective than the brute force attack with a factor of 2^d.

The main idea of the biclique attack can be defined as follows: In AES a number of keys $K[I, j]$ will be calculated in the key schedule function. At any time during encryption algorithm the state will have 2^d internal states S. The ciphertext C_i can now be seen as a function of a key $K[I, j]$ and a specific state S_i.[7, p.5]. The adversary forms a set of 2^{2d} keys from the key space and regards the block ciphertext (BC) as a combination of two sub ciphertexts where f follows g:

$$BC = f \circ g \tag{3}$$

The data transform of a ciphertext is constructed of two parts:

1. The adversary constructs a structure of 2^d ciphertext parts C_i and also 2^d intermediate states S_j in connection with the key group $K[I, j]$. Then a partial decryption of C_i results in S_j given $K[I, j]$.
2. The adversary uses an oracle to decrypt ciphertext C_i with the key K_{secret}. If K_{secret} is found in $K[I, j]$ the state S_j maps to the plaintext P_i which propose a key candidate [7, p.5].

2 Related work

As a protection against Differential Power Analysis attacks, Ghellar and Lubaszewski [10, p. 32] propose the addition of a mapping function to the beginning of the AES algorithm followed by an inverse mapping function as a final step of the algorithm. With 30 irreducible polynomials of degree 8 over GF(2) and 8 generator elements associated, 240 representation of GF(2^8) can be created. The proposed implementation adds a mapping function to the original AES algorithm and through the selection of representation, operation parameters are added to the SubBytes and MixColumn operations. Also the RoundKey is added a mapping before performing the AddRoundKey operation. In the end algorithm an inverse mapping is performed. The mapping conversion is based on the change of base in linear algebra. A $GF(2^8)$ element is multiplied by an 8x8 binary matrix producing a new representation of the $GF(2^8)$ element.

A new S-Box structure is proposed by Cui and Cao [12]. The S-Box construction of AES is generally considered weak, because the construction has a vulnerability of a simple algebraic expression [12, p.2]. The complexity is increased by creation of APA (Affine-Power-Affine) structure. In the original AES S-Box there are only n + 1 items at most in the algebraic expression of an affine transformation of $GF(n^p)$. With the APA structure the number of items is increased to 253 while the inverse S-Box keeps 255 items.

AES implementations can be placed in special dedicated processors or embedded RISC processors. Tillich and Groβschädl have been examined three possible solutions to prevent side-channel attacks [11] on AES. The first solution is to implement the security critical parts of the processor data path using DPA (Differential Power Attack) resistant logic style. The second solution is a strict software countermeasure using random pre-charging at instruction level. This solution has an increase in execution time, but the use of instruction set extensions helps the performance. The third solution is using a mask unit and is based on a combination of hardware and software solutions. The security zone in this solution is using a storage for the mask and a mask generator. The impact on performance is rather small.

3 TNAF-based Dynamic AES

The main idea in our approach is to modify parts of AES by taking advantage of Elliptic Curve Cryptography (ECC) used as a public key system and in this way address all the three types of attacks discussed above.

ECC can be implemented efficiently as Koblitz curves [8, p. 114] – also called anomalous binary curves. In this version normally a τ-adic non-adjacent form (TNAF) [8, p 116] is used in the ECC main algorithm. The TNAF function [8, p.117] converts a private key to a unique sequence with length l of $\{0, \pm1\}$ depending on the private key value. The TNAF function guarantees that the average density of nonzero digits is approximately 1/3 of the length l.

The dynamic TNAF-based AES main algorithm is mainly divided into two parts: a) TNAF-based key schedule for AES and: b) TNAF-based main algorithm for encryption and decryption. The main purpose of this approach is to remove the linearity of the key scheduling mechanism and the predictability after execution of a round in the main algorithm.

The ECC provides a point Q(x, y) on a valid curve. AES can take advantage of this to create a new key schedule as a part of a TNAF-based AES main algorithm. In this way the key would be substantially longer and the "industrial strength" will be improved, because the mix of a key schedule based on Q(x, y) will be decided at runtime – and not as pre-decided algorithm.

By using a mix of a key schedule based on Q(x, y) the key space is larger and provides the possibility of changing the actual used key dynamically during encryption and decryption.

If the actual combination of MixColumn, ShiftRows, SubBytes and AddRoundKey is decided at runtime the AES is not any longer foreseeable and attacks as described above will be impossible because they all rely on knowledge of the static algorithm as described in FIPS-197 [1]. Because the security is based on the function itself and not the static path in algorithm the security will be improved by the runtime decided function execution.

As the prerequisite a public key exchange has been done. If for instance the public key exchange is a normal Elliptic Curve Diffie-Hellman (ECDH)[8, p.171] key exchange, the participants A and B ends up with a common share in the form of a point Q(x, y). By using Secure Plain Diffie-Hellman (SPDH), the man-in-the-middle problem can even be eliminated [9].

3.1 Attack Countermeasures in τ-adic Dynamic AES

The dynamic TNAF-based solution will create the problem of a new side-channel attack since the TNAF sequence is a function of the private key k. If the TNAF sequence can be read by for instance measuring the power consumption, the private key k can be calculated. This will now be solved. A TNAF sequence could for instance be:

```
-1001010010000000-10-1010000-10010000000-1000-10000-1010-1001010-10000010101010
00010-101000-10-10100-10000-1000-10-1001010100000000-1010101000010000010-10-1000-
101000010-100001000-100010101010000010000100001000100001000010-101000-10000000-1001000-100
000000-100-10-1010-1000-10-100101010-10-100-1000-1001010100-10-1000000000-1000010
1001010101010000010-10010000100001010101_
```

From this it can be seen that the length between ± 1 and ± 1 vary from 1 zero up to 9 zeros. In all there are 318 digits $\{0, \pm 1\}$ with the following distribution: 220 zeros, 58 ones and 40 minus one.

Removing the trailing zeros will make it practically impossible to recover the original sequence.

Now the following algorithm can be applied:

Algorithm 1: Removing trailing zeros

INPUT: A byte array TNAF_RESULT = TNAF(k),

int counter = 0

OUTPUT: A byte array TNAF_TRAILING

For length of TNAF_RESULT do

    if TNAF_RESULT [i] equals 0

      TNAF_TRAILING [counter] = 0

      while TNAF_RESULT [i + 1] equals 0

        i = i + 1

  else

      TNAF_TRAILING [counter] = TNAF_RESULT [i]

    counter = counter + 1

return TNAF_TRAILING

It must be clear that the point Q on the elliptic curve must be validated before it is used. The domain parameters are public and well-known by the participants and thereby the specific curve type is known. The validation can be done by verifying that a point $Q \neq 8$ and also verifying that the point Q is on the curve by calculating for instance:

$$E_a : y^2 + xy = x^3 + ax^2 + b. \tag{4}$$

Algorithm 2: Validation of the ECC point

INPUT: A basepoint P(x, y), Domain parameters D

PARTICIPANTS: A

OUTPUT: Bool *IsValid*

A receives point P(x, y)

A calculates $y^2 + xy = x^3 + ax^2 + b$ according to D

if $y^2 + xy = x^3 + ax^2 + b$ equals 0

    return True

else

    return False

If the value False is received, a new base point must be chosen or a new calculation of point P(x, y) must be performed.

3.2 TNAF-based Key Schedule for AES

The dynamic TNAF-based key schedule for AES uses both the x and y coordinate of the common share Q(x, y). First the trailing zeros are removed from the TNAF values of Q(x, y) (algorithm 1) and the curve is validated (algorithm 2). The creation of round keys is depending on TNAF value of the x coordinate and the TNAF value of the y coordinate. Next issue is to place the keys for key schedule in a common array which is done as follows:

Algorithm 3: TNAF-based key schedule for AES

INPUT: A common share Q(x, y)

VARIABLES: int i, int RKx[], int RKy[], int TNAFX[], int TNAFY[]

FUNCTIONS: function CalcRoundKey, function TNAF, function MixKeys

OUTPUT: TNAF-based Key expansion array TK[], mixed set of TNAF values in MixTNAFValues[]

A calculates TNAF(Q.x) into TNAFX[]

A removes trailing zeros from TNAFX[] (algorithm1)

A calculates TNAF(Q.y) into TNAFY[]

A removes trailing zeros from TNAFY[] (algorithm1)

RKx[] = CalcRoundKey(Q.x)

RKy[] = CalcRoundKey(Q.y)

TK[] = MixKeys(RKx, RKy)

MixTNAFValues[] = MixKeys(TNAFX, TNAFY)   (see algorithm 4)

return TK, MixTNAFValues

The general TNAF algorithm guarantees that 2/3 of the TNAF values are zeros. As explained above, in order to prevent new side-channels attacks trailing zeros are removed from the TNAF sequence of Q(x, y). In this way the key schedule is strengthened and the side-channel attacks and biclique attacks mentioned above will be avoided.

The MixKeys function algorithm looks like this:

Algorithm 4: MixKeys function of mixing round keys based on Q(x, y) or the Q(x, y) based TNAF values

INPUT: int rounds, int w1[], int w2[]

VARIABLES: int index, int counter

OUTPUT: Reordered expansion key array or TNAF values in out []

   int counter = 0 from index= 0 to rounds

     out [counter++] = w1[index]

     out [counter++] = w2[index]

return out

In the case of mixing the expansion keys the output is placed in a new reordered array MixKeyRK and in the case of mixing the TNAF values, the output is placed in another reordered array MixTNAFValues.

Now that the sequence of keys has been placed in the array MixKeyRK, more flexibility can be added: It can be decided in the setup if the order in MixKeyRK should be from start or reverse. It can even be decided if the order is jumping after another pattern. The number of rounds is still supposed to follow the original number: For the 128 bits AES the number of rounds is 10, for the 192 bits AES the number of rounds are 12 and for the 256 bits AES the number of rounds are 14. The same flexibility can be added in case of MixTNAF Values.

3.3   TNAF-based Main Algorithm for Encryption and Decryption

The AES main algorithm can be further strengthen with a runtime decided mix of AES operations. In this way attacks as SPA (Simple Power Analysis) can be much harder to perform.

Since the ordinary AES is a simple combination of permutation and substitution the strength of the AES algorithm relies on the basic security of the mix of AES operations not the order of operation execution.

The TNAF sequences of the keys x and y created for the key schedule above is now used for the algorithm for encryption and decryption. The TNAF value decides at runtime the mix of the execution of the MixColumn, ShiftRows, SubBytes and AddRoundKey operations. Here follows the TNAF-based main encryption and decryption algorithm:

Algorithm 5: TNAF-based main encryption algorithm for AES

INPUT: Plaintext p, key schedule array MixKeyRK, a set of TNAF values in MixTNAF Values, Q(x, y)

VARIABLES: len$_{MixKey}$, state array, Boolean XYOrder

FUNCTIONS: function CalcXYOrder, MixColumn, ShiftRows, SubBytes and AddRoundKey

OUTPUT: Ciphertext C

p -> state

XYOrder = CalcXYOrder(Q(x, y)) (see algorithm 6)
if XYOrder is True
    from i = 0 to len$_{MixKey}$
      if MixTNAFValues [i] = 0
        ByteSub(state)
      if MixTNAFValues [i] = 1
        ShiftRows(State)
      if MixTNAFValues [i] = −1
      MixColumn(State)
        AddRoundKey(MixKeyRK [i]) (state)

if XYOrder is False
    from i = 0 to len$_{MixKey}$
      if MixTNAFValues [i] = 0
        ByteSub(State)
      if MixTNAFValues [i] = 1
        MixColumn(State)
      if MixTNAFValues [i] = −1
        ShiftRows(State)
      AddRoundKey(MixKeyRK [i]) (state)
    state -> C
return C

In the case of 128 bits AES Len_{MixKey} should have the size 30 (3 x 10).

Because the private key is based on the common share $Q(x, y)$ another security feature could be added to blur the calculation: A simple method to reverse the encryption and decryption order decided at run-time can be added:

Algorithm 6: Run-time decided (x, y) order CalcXYOrder

INPUT: Secure common key $Q(x, y,)$, int bitNumber

VARIABLES: BitString bitString

OUTPUT: Boolean bXyOrder

  bitString = Add(Q.x, Q.y)

  if bitString(bitNumber) = True

      bXYOrder = True

  else

      bXYOrder = False

return bXYOrder

From this it can be seen that the order of the key schedule calculation of $Q.x$ and $Q.y$ can vary dynamically with the subtraction (binary addition) of $Q.x$ and $Q.y$ and afterwards test a bit in the resulting bit string.

Now we define the decryption algorithm:

Algorithm 7: TNAF-based main decryption algorithm for AES

INPUT: Ciphertext C, key schedule array MixKeyRK, MixTNAFValues

VARIABLES: Len_{MixKey}, state (array)

FUNCTIONS: function CalcXYOrder, InvMixColumn, InvShiftRows, InvSubBytes and AddRoundKey

OUTPUT: Plaintext P

C -> State

XYOrder = CalcXYOrder(Q(x, y))

IF XYOrder is True

    from i= $Len_{MixKey}-1$ down to zero

      AddRoundKey(MixKeyRK [i]) (state)

      if MixTNAFValues [i] = 0

        InvByteSub(state)

      if MixTNAFValues [i] = 1

        InvShiftRows(state)

      if MixTNAFValues [i] = −1

        InvMixColumn(state)

if XYOrder is False

    from i = $Len_{MixKey}-1$ down to zero

    AddRoundKey(MixKeyRK [i]) (state)

      if MixTNAFValues [i] = 0

        InvByteSub(state)

      if MixTNAFValues [i] = 1

        InvMixColumn(state)

      if MixTNAFValues [i] = −1

        InvShiftRows(state)

    state -> *P*

return *P*

4 Implementation

The algorithms have been implemented in C++. The software contains implementation of a Koblitz ECC with TNAF and a standard AES implementation regarding the basic functions (AddRoundKey, MixColumn, ShiftRows, ByteSub plus the inverse functions).

    The ECC implementation is divided into three layers: A basic layer, a field layer and an ECC main algorithm layer. The implementation is on the basic layer using a BitString struct. The ECC part has been tested against the main formula (4).

The TNAF algorithm is implemented the following way [8, p.117]:

Algorithm 8: TNAF algorithm

INPUT: $k = r_0 + r_1\tau \in Z[\tau]$

OUTPUT: TNAF(k)

i = 0
while $r_0 \neq 0$ OR $r_1 \neq 0$ do
   if r_0 is odd then
         $u_i = 2 - (r_0 - 2\,r_1 \bmod 4)$
         $r_0 = r_0 - u_i$
   else
         $u_i = 0$
         $t = r_0$
         $r_0 = r_1 + \mu r_0/2$
         $r_1 = t/2$
         i = i + 1
return $(u_{i-1}, (u_{2-1}, \ldots, u_1, u_0)$

5 Tests and Results

Bob wants to send a message to *Alice*: "DYNAMIC AES". In order to send this secret message initially a normal Diffie-Hellman key exchange is performed between the two participants *Bob* and *Alice*.

First a base point Q (163 bits) is chosen:

X: 5c94eee8 de4e6d5e aa07d793 7bbc11ac fe13c053 2
Y: ccdaa3d9 0536d538 321f2e80 5d38ff58 89070fb0 2

The two participants *Bob* and *Alice* have each a private key:

Bob: 3456abcd 50567367 ab568676 67556316 000000aa 00000001

Alice: fe562343 00567766 ab568863 34556668 23673ab7 00000002

Having the base point Q *Bob* and *Alice* calculate a common (secret) share C using the Diffie-Hellmann algorithm:

X: f1cc692c 86527792 31d37422 cb346bdf fc76aef0 00000000

Y: beea1cf7 b59ff099 28852977 e726d75d b06beab5 00000003

The common secret share is then tested (algorithm 1) to be on the curve by placing it into the equation (4). This ends the normal Diffie-Hellman key exchange. The output $K(x, y)$ is now used as the symmetric key in AES.

The original AES algorithm uses a 4 x 4 byte state array as internal working array. Therefore one block of data to be encrypted or decrypted is of the size of 16 bytes. This array is reused in the TNAF based algorithm.

In the proposed TNAF based encryption algorithm the internal working process is:

1. Places the plaintext bytes in the state array
2. Perform key expansion of the private key (common share) (K.x and K.y)
3. Calculates TNAF values of the private key (common share) (K.x and K.y)
4. Removes trailing zeros from the TNAF sequences
5. Perform the MixKeys functions
6. Calculates the XY Order
7. According to the TNAF values from the MixKey function (reordered TNAF values) runs through SubBytes, MixColumn, ShiftRows and AddRoundKey functions

The key expansion is reused from the original AES algorithm. Example on calculating the TNAF values (before removing trailing zeros): From the common secret share the TNAF values and the number of values in $K.x$ and $K.y$. An example of the distribution $\{0, \pm1\}$ can be seen in Table 1.

Using the MixKey function the TNAF values from $K.x$ and $K.y$ are combined into one sequence of TNAF values with trailing zeros removed from TNAF $(K.x)$ and TNAF $(K.y)$.

Table 1   TNAF values

	0	1	−1	Total
K.x	208	58	53	319
K.y	207	50	64	321

In order to obtain the same minimum of function call (call to SubBytes, MixColumn, ShiftRows and AddRoundKey) as in the original AES algorithm, the number of calls for a 192 bit key is set to 44.

The XYOrder is in this case true. From this follows that the sequence is as follows:

1. SubBytes(state)
2. ShiftRows(state)
3. MixColumns(state)
4. AddRoundKey(state, key_schedule);

The output ciphertext is:

> 0x 59 92 63 21 C2 8A C1 5A DD FE C7 52 77 B3 0D 54

The decryption algorithm is still performing the key expansion and the MixKey function. Also the XYOrder function is called to decide the order. The decryption algorithm uses the original InverseSubBytes, InverseMixColumn and the InverseShiftRows functions from the original AES algorithm. The TNAF values are run through in reverse order. The functions above are called in the same order as in encryption. The AddRoundKey function is called first, so the calling sequence is as here:

1. AddRoundKey(state, RK [index]);
2. InverseSubBytes(state)
3. InverseShiftRows(state)
4. InverseMixColumns(state)

The output plaintext is "DYNAMIC AES".

5.1 Test Setup

In order to test the Dynamic AES a full implementation of ECC and Dynamic AES have setup. To give more precise time consumption for the specific parts of Dynamic AES the algorithms have been run through 10000 times each.

To avoid any caching in memory, the encryption algorithm has encrypted random generated data for each encryption block.

The dynamic AES will result in the same number of rounds as in the original AES in the main algorithm.

The TNAF-based AES along with ECC has been tested on an Intel(R) Core^{TM}2 Duo CPU @ 2.4 GHz.

5.2 Time Consumption Measurements

The proposed key scheduling mechanism is more time consuming, since two keys have to be scheduled and mixed combined with a calculation of two sets of TNAF values and a mix of keys. The time used for key scheduling has been measured in the implementation and the result is compared to the basic key scheduling function in standard AES as it can be seen in Table 2.

The proposed Dynamic TNAF-based AES is using the same basic operations as the original AES in the main algorithm.

The main algorithm has been tested against a standard AES implementation. Table 3 shows the time consumption results of encryption and decryption algorithms for Dynamic AES and standard AES.

We will now consider the expected throughput over a network using the values in tables 2 and 3. We assume that we want to transfer a file of 10 Mb data:

10 Mb = 10,240,000 bytes = 81,920,000 bits

The encryption and decryption time plus key schedule is calculated for a block of 16 bytes. The encryption operations can be done in 45.4501 sec and the decryption operations can be done in 46.901 sec. This means that the encryption flow is 1,802,416 bps and the decryption flow is 1,746,658 bps.

Table 2   Time Consumption Key Scheduling Mecahnism

	Dynamic AES Key Schedule	Standard AES Key Schedule
Time [mS]	10.094	0.0031

Table 3   Time Consumption Encryption /Decryption

	Dynamic AES Encryption	Standard AES Encryption	Dynamic AES Decryption	Standard AES Decryption
Time [mS]	0.071	0.044	0.072	0.045

6 Analysis of Results

Not surprisingly, the proposed key schedule algorithm, as it can be seen from Table 2, is significantly more time consumable than the original key scheduling algorithm of AES. However, normally this algorithm is only executed once per session. The reason for this overhead is first of all the calculation of TNAF values. Second, the trailing of zeros and third, the extended calculation of round keys which uses more time compared to the original AES key schedule operation. Also the last operations, the MixKey operation and calculation of the (x, y) order, will consume more time compared to the original algorithm.

The encryption and decryption algorithms are also more time consumable as it can be seen in Table 3, even though the original AES standard operations are used. This is caused by the additional logic to decide which standard AES operation is going to be performed at runtime. A future run-time optimization of the Dynamic AES main loop instruction sequence is expected to remove this problem.

When considering the transfer of a 10 Mb file over network using a Dynamic AES session, the peak performance is expected to be limited more by the network than by the algorithm for instance when considering HTTPS (and Dynamic AES extended version hereof) over the Internet. As mentioned, the Dynamic AES is well suited to be together with at public key system as for instance elliptic curve cryptography. As it can be seen above the network traffic time is not part of the test. If the network traffic time was included, the time consumption difference between the standard AES and the Dynamic AES would even be less important.

7 Conlusions and Discussions

The proposed Dynamic AES approach is based on the TNAF function used in Koblitz curves in ECC. It will solve the problem of side-channel attacks, related sub-key attacks and biclique attacks on AES. The proposed AES algorithm is adding a dynamic approach to the key schedule mechanism and the main algorithm of the original AES. The key scheduling algorithm of the original AES algorithm has been improved with the combined use of ECC. Any side-channel attack based on the use of TNAF algorithm is removed by removing trailing zeros. The private key is now longer and the linearity has been removed since two private keys belonging to the ECC algorithm can be mixed. The main algorithm of AES has been improved by adding dynamic behavior instead of a static run through. With this enhancement, the content of the state array in AES becomes unpredictable.

The TNAF-based Dynamic AES has been implemented in C++ along with ECDH based on Koblitz curves. The new algorithm will differ slightly in processing time primarily because the original AES key schedule is changed but also the additions in the main algorithm will have a minor cost in terms of performance.

In the future we will look into run-time code optimization of the Dynamic AES.

8 References

[1] FIPS Pub 197, NIST, November 26, 2001

[2] Jaon Daemen, Vincent Rijmen, The Rijndael Block Cipher, csrc.nist.gov, Sep. 1999

[3] Joseph Bonneau, Side-Channel Cryptoanalysis (Research students' Lectures), University of Cambridge Computer Laboratory, May 4, 2010.

[4] Dag Arne Osvik, Adi Shamir and Eran Tromer, Cache Attacks and Countermeasures: the Case of AES, osvik.no / Department of Computer Science and Applied Mathematics, Weizmann Institute of Science, Rehovot 76100, Israel, 2005

[5] Joseph Bonneau and Ilya Mironov, Cache-Collision Timing Attacks Against AES, Computer Science Department, Stanford University and Microsoft Research, Silicon Valley Campus, 2006

[6] Alex Biryukow and Dmitry Khovratowich. Related-Key Cryptanalysis of the Full AES-192 and AES-256, p.1–18, ASIACRYPT 2009

[7] Andrey Bogdanov, Dmitry Khovratovich, and Christian Rechberger, K.U. Leuven, Belgium; Microsoft Research Redmond, USA; ENS Paris and Chaire France Telecom, France, Biclique Cryptanalysis of the full AES, ASIACRYPT'11, August 31, 2011

[8] Hankerson, Menezes and Vanstone, "Guide to Elliptic Curve Cryptography", Springer, 2004.

[9] Henrik Tange, Birger Andersen, Secure Plain Diffie-Hellman algorithm, Journal of Cyber Security and Mobility, 2012

[10] Felipe Ghellar, Marcelo Soares Lubaszewski, A Novel AES Cryptographic Core Highly Resistant to Differential Power Analysis Attcks, Jorunal Integrated Circuits and Systems, 2009

[11] Stefan Tillich, Johann Groβschädl, Power Analysis Resistent AES Implementation with Instruction Set Extensions, LNCS 4727, 2007

[12] Lingguo Cui, Yuanda Cao, A New S-Box Structure Named Affine-Power-Affine, ICIC International, 2007

Biographies

Henrik Tange received the B.Eng (export engineer) from the Copenhagen University College of Engineering in 1999 and the M.Sc. in Communication Network specializing in Security from Aalborg University in 2009. Since 2009 he has been a PhD student at Aalborg University. Since 2000 he has been teaching at Copenhagen University College of Engineering which merged into Technical University of Denmark.

Birger Andersen is a professor at Technical University of Denmark, Copenhagen, Denmark, and director of Center for Wireless Systems and Applications (CWSA). He received his M.Sc. in computer science in 1988 from University of Copenhagen, Denmark, and his Ph.D. in computer science in 1992 from University of Copenhagen. He was an assistant professor at University of

Copenhagen, a visiting professor at Universität Kaiserslautern, Germany, and an associate professor at Aalborg University. Later he joined the IT department of Copenhagen Business School, Denmark, and finally Copenhagen University College of Engineering which merged into Technical University of Denmark. He is currently involved in research in wireless systems with a focus at security

Performance Evaluation of Beamspace MIMO Systems with Channel Estimation in Realistic Environments

Konstantinos Maliatsos, Panagiotis N. Vasileiou and
Athanasios G. Kanatas

Department of Digital Systems, University of Piraeus, Greece,
E-mail: {kmaliat;panvas;kanatas}@unipi.gr

Received 27 October 2013; Accepted 14 December 2013;
Publication 23 January 2014

Abstract

Beamspace MIMO (BS-MIMO) systems have been recently proposed as a means to address the two key weaknesses of conventional MIMO systems: the antenna size and the need for multiple RF chains. BS-MIMO transmission is supported by the Electronically Steerable Passive Array Radiators (ESPAR) with a single active and multiple parasitic elements. The main objective is to develop efficient MIMO multiplexing schemes that use only one Radio Frequency (RF) chain and simultaneously maintain extremely small antenna size. Moreover, the recent research results have shown that BS-MIMO systems have increased multiplexing and beamforming capabilities and for small antenna sizes clearly outperform equivalent conventional systems in terms of system capacity. Nevertheless, research on BS-MIMO has been focusing on the study of the ESPAR antenna properties that facilitate beamspace transmission and the theoretical analysis of the ergodic capacity provided by the aerial degrees of freedom (aDoF) of the beamspace channel. Recently, the first steps have been made to design and evaluate practical BS-MIMO systems. This paper presents extended results of the first attempt to design practical and realistic BS-MIMO transmission and reception schemes and it specifically focuses on channel estimation techniques for BS-MIMO systems with adaptive pattern reconfiguration. Adaptation of the basic least-squares (LS) and minimum mean squared error (MMSE) estimators for the beamspace radio channels is performed and the algorithms are incorporated in an adaptive Singular Value Decomposition

Journal of Cyber Security, Vol. 2 No. 3 & 4, 265–290.
doi: 10.13052/jcsm2245-1439.234

(SVD)-based system. Finally, fundamental results extracted by the developed beamspace link level simulator are presented in order to evaluate and compare BS-MIMO with equivalent conventional MIMO systems.

Keywords: beamspace, MIMO, BS-MIMO, channel estimation, ESPAR, link level evaluation.

1 Introduction

Modern and future radio systems consider Multiple Input Multiple Output (MIMO) transmission techniques as the means to significantly improve spectral efficiency and system capacity. However, the conventional MIMO transceivers require the use of multiple Radio-Frequency (RF) chains in order to feed the elements of an antenna array causing a significant increase of the development cost. Moreover, in order to obtain adequate spatial multiplexing and/or beamforming properties, the antenna array of the MIMO systems should have large physical dimension. The increased antenna size will ensure decoupled elements and low spatial correlation at the transmitter/receiver.

The Single RF [1] BS-MIMO systems [2] transfer the MIMO operation from the antenna elements to beamspace. Instead of using the voltages or currents on the antenna elements to directly carry the transmitted symbols, the data streams are mapped onto a selected set of radiation patterns that constitute an orthonormal basis in the beamspace domain [2], [3] and through beamspace multiplexing data are sent simultaneously to the wireless channel. In contrast to the conventional technique, BS-MIMO multiplexing and beamforming capabilities are favoured by the use of compact antenna arrays with closely spaced elements, since adaptive pattern reconfiguration is achieved with the control of the currents that are induced on the parasitic elements from the active element. The Electronically Steerable Passive Array Radiators (ESPAR) [4] with one active and multiple parasitic elements is an excellent choice able to support implementable BS-MIMO transceivers.

The multiplexing capabilities in the beamspace domain are quantified by the aerial Degrees of Freedom (aDoF). In [5], it was proved that based on the ESPAR antenna properties and assuming an ideally rich scattering radio channel, the ESPAR-based BS-MIMO system is able to simultaneously transmit M orthogonal data streams, where M is the total number of antenna elements (i.e. 1 active and $M - 1$ parasitic elements). In order to maximize the exploitation of the available system capacity for realistic beamspace channels, an adaptive pattern reconfiguration scheme that is based on Singular Value Decomposition

(SVD) was proposed in [6]. Perfect channel state information in both transmitter and receiver was assumed introducing an adaptation scheme similar with the SVD precoding in the conventional MIMO. Despite the fact that the results on pattern reconfiguration with the use of ESPAR antennas [7] are quite promising and that the achievable ergodic capacity in the beamspace domain exceeds the equivalent results from conventional MIMO architectures, no attempt to evaluate the link level performance of BS-MIMO transceivers with the use of realistic reception algorithms and propagation conditions was made until the work presented in [8].

This paper presents and extends the results of [8] provided for a 3-element ESPAR system with the use of the more efficient and flexible 5-element ESPAR antenna. An implementable beamspace transceiver chain is proposed and a link-level simulator is developed. The main focus is given on the formulation of channel estimation algorithms for BS-MIMO. The channel estimation blocks together with the adaptive pattern reconfiguration procedures are incorporated in the transmission chain. The beamspace transmission system model is presented and analysed in Sec. 2. Then, the estimation algorithms (least squares - LS and minimum mean squared error - MMSE) are formulated for the beamspace and the optimal training patterns are identified in Sec. 3. Moreover, the concept of the interfering patterns is introduced. In Sec. 4, the developed link level simulator is presented and extended simulation is performed in order to evaluate the BS-MIMO performance compared to an equivalent conventional MIMO system. The conclusions of Sec. 5 confirm that BS-MIMO has the potential to provide significant advantages over the conventional systems when compact antenna arrays are applied, and some main issues are identified that should be addressed in order to implement a fully functional and efficient BS-MIMO transceiver.

Notation: Bold lowercase letters represent vectors and bold capital letters matrices. $(\cdot)^{H}$ is used to describe the matrix conjugate transpose, while $[\cdot]_{i,j}$ is used to describe the (i,j)-element of the included matrix. $\|\cdot\|_{F}$ is the Frobenius norm. Function Tr calculates the trace of a matrix. The bs superscripts or subscripts are used to illustrate that a specific matrix or quantity is provided in the beamspace.

2 BS MIMO System Model and Pattern Adaptation

The common input-output relationship for conventional MIMO systems through a flat fading channel with Additive White Gaussian Noise (AWGN) is given by:

$$y = Hx + n \tag{1}$$

where **y**, **x** and **n** are the received (Rx) signal, transmitted (Tx) signal and noise vectors respectively, while **H** is the channel matrix. In conventional MIMO, all the antenna elements are active terminated to a 50 Ohm load. The data signal **x** physically represent the voltages or equivalently the currents that flow on the antenna elements. The multiplexing capability of the MIMO channel, i.e. its ability to support parallel transmission of data stream is represented by the available Degrees of Freedom (DoF). For a given channel, the DoF of the system are calculated through the SVD of the channel matrix. For small sized arrays the DoF are significantly reduced due to the high values of spatial correlation between the antenna elements of the transmitter and/or the receiver, that also increases the correlation between the elements of **H**. The Tx spatial correlation for a Uniform Linear Array (ULA) antenna is given by [9]:

$$[\mathbf{R}_{hh}]_{i,j} = \left[E\left(\mathbf{h}_m \mathbf{h}_m^{\mathrm{H}}\right)\right]_{i,j} = \frac{1}{M_T^2} J_0\left(\frac{2\pi d\,|i-j|}{\lambda}\right) \tag{2}$$

where $\mathbf{h}_m^H$ the m−th line of **H**, J_0 the zeroth-order Bessel function, d the interelement distance, λ the wavelength and M_T (M_R) the Tx (Rx) antenna elements. In this paper a 3-element and a 5-element ULA with interelement distance $d = \lambda/16$ is considered for the equivalent conventional MIMO system. Assuming a normalized MIMO matrix **H** ($\|\mathbf{H}\|_F^2 = 1$) and using (2) for the 5-element ULA:

$$[\mathbf{R_{hh}}]_{1,1\ldots5} = \begin{bmatrix} 1 & 0.96 & 0.85 & 0.68 & 0.47 \end{bmatrix}/25 \tag{3}$$

It is clear that heavy correlation among **H** elements is introduced. Thus, it is statistically expected that DoF < 5 reducing the overall system capacity. For realistic channels that vary considerably from the ideally rich scattering environment, it is possible that despite the fact that 3 or 5 antennas are used, the MIMO system will degenerate to Single Input Single Output (SISO) i.e. DoF $= 1$

In BS-MIMO, multiplexing is performed in the beamspace domain. The information for all the parallel streams in a given instance is carried by the radiated pattern that is shaped using a parasitic array with a single active element. Initially, a set of orthonormal basis patterns is defined. Beamspace multiplexing is achieved with the transmission of the pattern that is resulted by the linear combination of the basis patterns with the information symbols. The

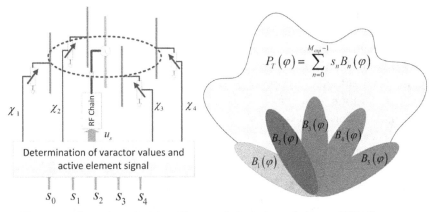

Figure 1 Illustration of BS-MIMO transmission using a 5-element ESPAR antenna

Tx pattern that carries the symbol set at a specific time sample n is provided by [5]:

$$P(\varphi, n) = [B_0(\varphi) \ldots B_{M_T-1}(\varphi)] \mathbf{x}_{bs}(n) \qquad (4)$$

where B_i is the i−th basis pattern and $\mathbf{x}_{bs}(n)$ the data vector at time sample n. For simplicity, azimuth plane propagation only is assumed (angle ϕ).

The ESPAR antenna should be able to adapt the Tx pattern per symbol period using one active and $(M_T - 1)$ parasitic elements. BS-MIMO transmission is implemented, if the ESPAR antenna transmits the desired pattern $P(\varphi, n)$. This is achieved with proper readjustment of the variable reactance values (varactors) of the parasitic element loads $\chi_{i=1 \ldots M_T-1}$. In order to determine the varactor values that produce the desired pattern, the following relationship is used [2]:

$$P(\varphi, n) = \mathbf{a}(\varphi)^T \mathbf{i} = \mathbf{a}(\varphi)^T v_s(n) (\mathbf{Z} + \mathbf{X}(n))^{-1} \begin{bmatrix} 1 & 0 & \ldots & 0 \end{bmatrix}^T \quad (5)$$

where $\mathbf{Z}$ is the $M_T \times M_T$ matrix of the mutual coupling among the antenna elements, $\mathbf{a}(\varphi)$ the array manifold vector, $\mathbf{i}$ the vector of the currents that flow on the active and parasitic elements, $v_s(n)$ the excitation signal of the active element and $\mathbf{X}(n)$ the diagonal matrix that contains the vararctor values $j\chi_{i=1 \ldots M_T-1}$ of the parasitic elements during the n−th symbol. The specific algebraic representation assumes that the active element is placed in the first line of matrix $\mathbf{X}(n)$, thus $[\mathbf{X}]_{1,1} = 50$ Ohm since the active element is terminated to a 50 Ohm resistance. Calculation of desired load values can be performed by (5) with the use of a stohastic [7] or a genetic [10] algorithm. Assuming for simplicity that $M_T = M_R = M$, the reception is implemented with the sequential alternation of the basis patterns during a symbol period (with an M-times oversampling rate).

The basis patterns are selected to form an orthonormal set of functions for two main reasons. Firstly, in [2] a procedure that allows the use of the common MIMO input-output relationship (1) to also describe BS-MIMO transmission is presented. Assuming a channel with K discrete scatterers and $\mathbf{H}_g$ is the $K \times K$ complex, diagonal scattering matrix, then the beamspace transmission can be modelled as:

$$\mathbf{y}_{bs} = \mathbf{H}_{bs}\mathbf{x}_{bs} + \mathbf{n}_{bs} = \mathbf{B}_R^H\mathbf{H}_g\mathbf{B}_T\mathbf{x}_{bs} + \mathbf{n}_{bs} \tag{6}$$

where the $K \times M$ matrices $\mathbf{B}_{T/R}$ contain the values of the M basis patterns in transmitter and receiver towards the direction of the scatterers (Angles of Departure in the Tx and Angles of Arival in the Rx). It is emphasized that in beamspace $\mathbf{x}_{bs}$ and $\mathbf{y}_{bs}$ are not directly related to the element currents but their relationship with the currents is provided by (5). Thus, $\mathbf{H}_{bs}$ and $\mathbf{H}$ have completely different structures due to the performed linear transformations [2]. Investigation of (6) leads to the conclusion that conventional MIMO algorithms can also be used for BS-MIMO systems with no significant changes. Secondly, it is proved that the (Tx or Rx) correlation matrix for each MISO cylindrically isotropic 2-D channel that describes the ideally rich scattering propagation environment is given by:

$$\mathbf{R}_{hh} = \left(1/M^2\right)\mathbf{I}_M \tag{7}$$

where $\mathbf{I}_M$ is the $M-$identity matrix. Thus, the selection of the orthonormal basis patterns leads to the best possible correlation properties for rich scattering environments. In [6], it was shown that the system maintains its strong correlation properties in realistic radio channels. Therefore, it is expected that matrix $\mathbf{H}_{bs}$ will provide aDoF that are statistically expected to approach M *regardless of the interelement distance d* and thus the system will provide better support for the transmission of multiple data streams.

The aDoF can be determined with the SVD of the channel matrix. Moreover, the result of the SVD can be used to develop and implement an adaptive pattern reconfiguration scheme to maximize capacity and channel efficiency. In [6], a pattern adaptation rule, similar with the conventional MIMO SVD-based precoding, is proposed. If $\mathbf{H}_{bs} = \mathbf{U\Sigma V}^H$, then it is proved that the beamspace channel can be exploited optimally with the use of adapted basis patterns at Tx and Rx. More specifically, the new orthonormal bases are given by:

$$\bar{\mathbf{B}}_R = \mathbf{B}_R\mathbf{U}_{aDoF}, \quad \bar{\mathbf{B}}_T = \mathbf{B}_T\mathbf{V}_{aDoF} \tag{8}$$

The subscript of $\mathbf{U}$ and $\mathbf{V}$ defines that the first aDoF columns of the matrices are used, reducing the used basis patterns, if necessary, with respect to the supported aDoF. In order to practically implement and evaluate the pattern adaptation scheme, it is necessary to extract accurate channel estimates. Moreover optimization of the capacity can be performed with proper power allocation among the patterns with the use of the waterfilling algorithm. More specifically, the power per pattern is assigned with the following formula:

$$p_i = \left(\mu - \frac{\sigma_n^2}{\sigma_i^2} \right)^+ \tag{9}$$

where σ_n^2 is the noise variance, σ_i is the i-th singular value of Σ and μ is a constant that depends on the total available transmitted power i.e. $P_{tx} = \sum_{i=0}^{M-1} \left(\mu - \frac{\sigma_n^2}{\sigma_i^2} \right)^+$. Superscipt $+$ indicates that the expression is valid for positive values of the argument (otherwise $p_i = 0$). If $\mathbf{W} = \text{diag} \begin{bmatrix} p_1 & \cdots & p_{aDoF} \end{bmatrix}$ is the power allocation matrix, then the transmitted pattern towards the scatterers is given by:

$$\mathbf{p}_T = \mathbf{B}_T \mathbf{V}_{\text{aDoF}} \sqrt{\mathbf{W}} \mathbf{x}_{bs} \tag{10}$$

This paper assumes that sets of 3-element or 5-element ESPAR antennas are used in Tx and Rx in order to evaluate BS-MIMO performance, while a 3-element or 5-element ULA with the same interelement distance $d = \lambda/16$ is assumed for the equivalent conventional MIMO system. The initial channel-unaware orthonormal basis patterns can be analytically extracted with the use of a proper othogonalization algorithm e.g. Gram-Schmidt. The analytical expressions for the 5-element ESPAR antenna can be found in [5] and for the 3-element ESPAR in [8].

3 Channel Estimation in BS-MIMO

The equivalence between the algebraic representations of BS-MIMO and conventional MIMO systems can be exploited in order to transfer algorithms from the conventional systems [11] to the beamspace domain. Moreover, the presented channel estimation algorithms can be used for wideband, frequency selective channels. The delay is introduced as the third dimension and therefore the channel is transformed in a 3-D matrix $\mathbf{H}_{bs}(k)$ with $k = 0, 1..L$, where L is the maximum expected excess delay. In beamspace systems, each element of the channel matrix represents the complex gain between beams, that are expressed in the angular domain through the basis patterns. Frequency selectivity in a beamspace transmission is caused by energy from previously transmitted patterns that spreads in time due to the different propagation paths and corresponding delays of the radio channel. The delayed arrival of signals at the receiver causes InterSymbol Interference (ISI). Wideband BS-MIMO analysis is a subject for future work, however the presented channel estimation algorithms can be applied to frequency selective channels. The specific channel estimation algorithms assume that a training sequence (known to the Rx) is used. The training sequence is carried by the initial channel-unaware basis patterns forming the *training patterns*. Given the fact that $M^2(L+1)$ parameters are estimated, the training sequence length N_T for each MISO channel should be greater than $M(L+1)$ samples ($MN_T \geq M^2(L+2)$).

The algorithms for conventional MIMO systems in [11] decompose the channel estimation procedure in MISO channels. Îd'his decomposition is valid when uncorrelated spatial reception is assumed. For small-sized antennas the assumption of uncorrelated reception is false for conventional MIMO as shown in (3), however it is 100% accurate for BS-MIMO systems. Thus, the channel for each received beam $\mathbf{h}_{bs,i}^{H}(k)$ (each line of the channel matrix) can be extracted separately to form the estimate of $\mathbf{H}_{bs}(k)$. In order to mathematically define the estimators, the 3-D matrices are rearranged and the following reshaped vector of parameters is defined for the i-th MISO channel:

$$\mathbf{g}_i = \left[h_{i,0}^{bs}(0) ... h_{i,0}^{bs}(L) |...| h_{i,M-1}^{bs}(0) ... h_{i,M-1}^{bs}(L) \right] \qquad (11)$$

The observation vector for each MISO channel used by the estimators is:

$$\mathbf{z}_i = \left[\; y_{bs,i}(L) \quad \cdots \quad y_{bs,i}(N_T - 1) \; \right]^T \qquad (12)$$

It is observed that the estimators ignore the L first incoming samples that are affected from preceding symbols not contained in the training sequence.

3.1 Linear Least Squares (LS)

In compliance with the shape of the vectors defined in (11) and (12), a $((N_T - L) \times M(L+1))$ matrix is defined that contains the training sequences carried by the M basis patterns:

$$\mathbf{X} = \left[\; \mathbf{X}_0 \quad \cdots \quad \mathbf{X}_{M-1} \; \right],$$

$$\mathbf{X}_i = \begin{bmatrix} x_{bs,i}(L) & \cdots & x_{bs,i}(0) \\ x_{bs,i}(L+1) & & x_{bs,i}(1) \\ \vdots & & \vdots \\ x_{bs,i}(N_T-1) & \cdots & x_{bs,i}(N_T-L-1) \end{bmatrix} \tag{13}$$

The LS estimate is then given by:

$$\hat{\mathbf{g}}_{i,LS} = \left[\mathbf{X}^H\mathbf{X}\right]^{-1}\mathbf{X}^H\mathbf{z}_i \tag{14}$$

The resulted MISO mean squared estimation error is given by:

$$\varepsilon_{LS,MISO} = \sigma_n^2 \mathrm{Tr}\left[\left(\mathbf{X}^H\mathbf{X}\right)^{-1}\right] \tag{15}$$

where σ_n^2 is the noise variance. For zero-mean uncorrelated errors with equal variances (since noise is generally assumed equal for all MISO channels), the LS estimator is the best linear unbiased estimator.

3.2 Minimum Mean Square Error (MMSE)

The mean squared error of the estimation is minimized with the MMSE that uses a priori information:

$$\hat{\mathbf{g}}_{i,MMSE} = \frac{1}{\sigma_n^2}\left[\mathbf{R}_{hh}^{-1} + \frac{\mathbf{X}^H\mathbf{X}}{\sigma_n^2}\right]^{-1}\mathbf{X}^H\mathbf{z}_i \tag{16}$$

A priori information is contained in the correlation matrix $\mathbf{R}_{hh}$ of vector $\mathbf{g}_i$, i.e. the correlation matrix of the i-th wideband MISO channel. For $L = 0$ (flat channel) and for rich scattering environments [13], the correlation matrix is approximated by (7) since uncorrelated beamspace transmission is achieved. In case of a wideband channel, each diagonal element is replaced by an $(L+1) \times (L+1)$ submatrix which is the correlation matrix for each beam in the delay domain. Assuming channels with uncorrelated scattering, $\mathbf{R}_{sub}$ will contain in its diagonal an estimate of the channel power delay profile per beam. If this information is unavailable, various empirical rules can be applied:

$$\mathbf{R}_{sub} = \underbrace{\begin{bmatrix} 1/(L+1) & & 0 \\ & \ddots & \\ 0 & & 1/(L+1) \end{bmatrix}}_{\text{Large Delay Spread}} \text{or} \underbrace{\begin{bmatrix} 1 & & 0 \\ & 0 & \\ 0 & & \ddots \end{bmatrix}}_{\text{Small Delay Spread}} \tag{17}$$

In these expressions the matrices are normalized to unit power. The first case assumes uniform power delay profile and it should be used for channels with large delay spread. The second case initially assumes flat fading. However, the second approach results in a non-invertible matrix. In order to use this approach, low-power noise should be added to the zero-valued elements of the matrix diagonal. Although MMSE performs well using the aforementioned empirical rules, in slow varying channels better results can be achieved when the previous channel estimate is used to approximate the correlation matrix of the current channel. Thus,

$$\hat{\mathbf{R}}_{hh} = \hat{\mathbf{g}}_{i,\mathrm{old}}\hat{\mathbf{g}}_{i,\mathrm{old}}^{H} \tag{18}$$

However, the specific approximation has been proven sensitive to phase noise and it requires very low mobility. Better and stable performance is achieved with the use of the previous channel estimate as an approximation of the power delay profile, assuming uncorrelated scattering. In this case, $\hat{\mathbf{R}}_{hh}$ can be written as:

$$\left[\hat{\mathbf{R}}_{hh}\right]_{i,j} = \begin{cases} \left|\left[\hat{\mathbf{g}}_{i,\mathrm{old}}\right]_{i}\right|^{2}, & \text{if } i = j \\ 0, & \text{if } i \neq j \end{cases} \tag{19}$$

The procedure can be initialized using an empirical rule from (17).

The theoretically calculated estimated error is given by:

$$\varepsilon_{\mathrm{MISO}}^{\mathrm{MMSE}} = \mathrm{Tr}\left[\left(\mathbf{R}_{hh}^{-1} + \frac{\mathbf{X}^{H}\mathbf{X}}{\sigma_{n}^{2}}\right)^{-1}\right] = \mathrm{Tr}\left[\underbrace{\left(M^{2}\mathbf{I}_{M} + \frac{\mathbf{X}^{H}\mathbf{X}}{\sigma_{n}^{2}}\right)^{-1}}_{\text{flat,rich scattering}}\right] \tag{20}$$

where the result for a rich flat normalized channel is also provided.

3.3 Optimal Training Patterns

The training patterns are formed by sequential N_T linear combinations of the basis patterns with a properly selected training sequence. Since both LS and MMSE estimators depend on the inverse of the product $\mathbf{X}^{H}\mathbf{X}$, it is proved with the use of Lagrange multipliers that for given power of the training sequence, the errors are minimized when [18]:

$$\mathbf{X}^{H}\mathbf{X} = \begin{cases} \lambda_{LS}\mathbf{I}_{(L+1)M}, & \text{for LS} \\ \lambda_{MMSE}\mathbf{I}_{(L+1)M} - \mathbf{R}_{hh}^{-1}, & \text{for MMSE} \end{cases} \tag{21}$$

where the constants λ_{LS} and λ_{MMSE} are selected to satisfy the power constraints of the training sequence. For a flat, rich-scattering channel, or a wideband channel with large delay spread, the matrix $\mathbf{R}_{hh}$ is a scaled identity matrix and thus the LS and MMSE estimators have a common minimum. In this case, since $\mathbf{X}$ is by definition a Toeplitz matrix, a training sequence that diagonalizes $\mathbf{X}^H\mathbf{X}$ and minimizes the estimation error is the Frank-Zadoff-Chu (FZC) sequence [11]. The sequence is placed in the 1st column or row of $\mathbf{X}$ depending on which dimension is greater. The following columns/rows are filled with circular rotation of the previous column/row by an element. Thus, the FZC sequence size is either $N_T - L$ or $M(L + 1)$. Since the sequence size must exceed the number of estimated parameters, an FZC sequence of $N_T - L$ length is assumed. The FZC sequence is not a sequence of modulated symbols (e.g. using the QAM modulation), however it is mapped to the initial channel-unaware basis patterns with the same method as the data streams, forming the optimal training patterns. It is noted that channel estimation should always be made with the use of the channel-unaware patterns. The use of channel adapted patterns during the estimation stage may lead to false detection of the available aDoF, since the radio channel continuously changes in time. The optimal training patterns are given by:

$$P_{\text{train}}(\varphi, k) = \sum_{i=0}^{M-1} \mathbf{B}_i(\varphi) f(k - (i + 1)L), \quad k = 0...N_T - 1$$

$$\text{where: } f(k) = \begin{cases} e^{\frac{j\pi Qk^2}{N_T - L}}, & \text{if } N_T - L \text{ even} \\ e^{\frac{j\pi Qk(k+1)}{N_T - L}}, & \text{if } N_T - L \text{ odd} \end{cases} \tag{22}$$

where Q is a constant coprime to $N_T - L$.

If $\mathbf{R}_{hh}$ is not a scalable identity matrix, the FZC sequences are not optimal for the MMSE estimator. From [17], it can be proved that the optimal training sequences can be found by linear transformation of the FZC. If $\mathbf{R}_{hh} = \mathbf{Q}\mathbf{\Lambda}\mathbf{Q}^H$ is the eigenvalue decomposition of the hermitian matrix $\mathbf{R}_{hh}$ and since $\mathbf{Q}$ is an orthogonal matrix, then:

$$\mathbf{Q}^H\mathbf{X}^H\mathbf{X}\mathbf{Q} = \lambda\mathbf{I}_{(L+1)M} - \mathbf{\Lambda}^{-1} \tag{23}$$

which is minimized when $\mathbf{Q}^H\mathbf{X}^H\mathbf{X}\mathbf{Q}$ is a diagonal matrix and based on the previous analysis, when $\mathbf{Q}^H\mathbf{X}^H\mathbf{X}\mathbf{Q} + \mathbf{\Lambda}^{-1}$ is a FZC sequence.

In Fig. 2 and Fig. 3, mean squared estimation error results for a 5-element and a 3-element ESPAR antenna BS-MIMO system respectively are presented. The interelement distance in both scenarios was $d = \lambda/16$ and the WINNER II channel model [14] was used after proper modifications in

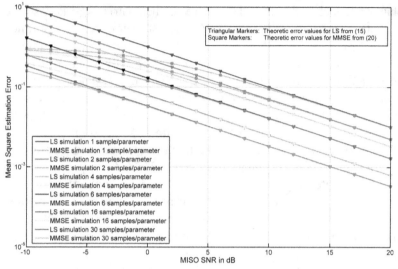

Figure 2   : Channel Estimation Error vs SNR in BS-MIMO for flat fading B2 WINNER channels and various training sequence lengths using a 5-element ESPAR antenna

order to support beamspace transmission [13]. The WINNER radio channels were downsampled for narrowband transmission (flat fading channels). The estimators had to determine $M^2 = 25$ or 9 complex gains of $\mathbf{H}_{bs}$. Training was performed with FZC sequences. The term samples/parameter is used as a measure of the training sequence length. If x training samples per parameter are selected, then the training sequence length per MISO channel is given by $N_T = Mx$. Evaluation was performed with 10,000 flat 'B2' (bad urban micro-cell environment) channels vs. SNR for various N_T values. For the MMSE estimator, the correlation matrix was set according to $\mathbf{R}_{hh} = \left(1/M^2\right)\mathbf{I}_M$. The simulation results (lines) and the theoretic results (markers) from (15) and (20) were identical, which also means that 'B2' and ideally rich scattering channels behave similarly in beamspace, as far as the estimation procedure is concerned. MMSE and LS results are practically equal for high SNR, while for low SNR MMSE performs better. The SNR value, where MMSE overcomes the LS estimator depends on the length of the training sequence. The investigation of the two Figures shows that the error values for e.g. 4 training samples/parameter are lower in the 3-element ESPAR scenario. This happens because for each MISO channel, the error is calculated as the sum of M error terms from the M transmitted beams. Since for the 5-element ESPAR M is greater, the error also increases. Accurate estimation at low

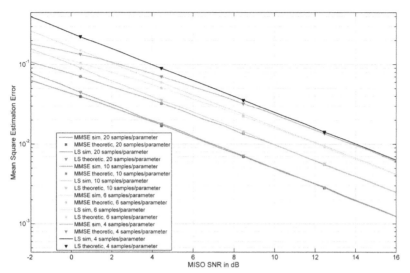

Figure 3   : Channel Estimation Error vs SNR in BS-MIMO for flat fading B2 WINNER channels and various training sequence lengths using a 3-element ESPAR antenna

SNR requires long training sequences. For example 30 samples/parameter produce error approximately equal to $\varepsilon = 0.02$ @5dB SNR for 5-element ESPAR, while the same error value is achieved with 20 symbols/parameter for the 3-element ESPAR.

The next step is to test the performance of the estimation algorithms in wideband beamspace channels. The previous 5-element antenna setup was considered. However, in this case the BS-MIMO system was assumed to occupy 1.5 MHz of bandwidth, and thus it cannot be considered narrowband. The training sequence length was 400 samples. Each signal frame contained 2,900 symbols (400 training samples + 2,500 data symbols) and the receiver was considered mobile (pedestrian user with velocity $3m/sec$). 2,000 'B2' WINNER channels were produced and the transmission of 30 frames of data was simulated. The WINNER channels were downsampled in the signal bandwidth (1.5 MHz). The maximum excess delay under these conditions was $L + 1 = 5$. Therefore, the algorithms should estimate $M^2(L + 1) = 125$ parameters. In Fig. 4, the results of the evaluation are presented for various approximations of $\mathbf{R}_{hh}$. It is seen that the MMSE algorithm despite the approximations, performs better than the LS. For the specific channels, the uniform power delay profile assumption (17) performs worse than the flat channel approximation (with addition of noise). The generic Cluster Delay

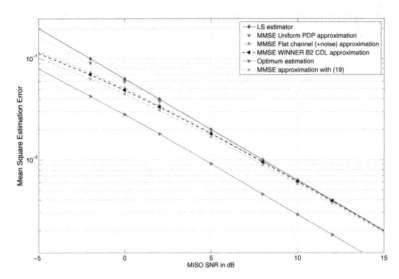

Figure 4 : Wideband Channel Estimation Error vs SNR in BS-MIMO for B2 WINNER channels ($L = 4$) with training sequence length = 400 samples using a 5-element ESPAR antenna

Line (CDL) profile of the WINNER 'B2' channels [20] was also used in the diagonal of $\mathbf{R}_{hh}$ and the estimation algorithm performance was slightly improved. Finally the approximation of (19) was evaluated. During the first frame, the flat approximation was used and then for the 29 succeeding frames, the previous estimation was used. It is noted that the receiver is mobile, thus the channel response varies from frame to frame. The estimation error was further reduced. For comparison reasons, an ideal case is also presented in Fig.4 where the MMSE operates with $\widehat{\mathbf{R}}_{hh} = \mathbf{g}_i\mathbf{g}_i^{\mathrm{H}}$, which means that the channel is actually known to the estimator.

3.4 Use of the Estimate for Pattern Reconfiguration

At the end of the estimation procedure, the channel matrix estimate $\hat{\mathbf{H}}_{bs}$ is available and it can be used for adaptive basis pattern reconfiguration according to the scheme proposed in [6]. The first step is to perform SVD decomposition $\hat{\mathbf{H}}_{bs} = \hat{\mathbf{U}}\hat{\mathbf{\Sigma}}\hat{\mathbf{V}}^{\mathrm{H}}$. The overall estimation error is divided in the 3 matrices, however the errors of $\hat{\mathbf{U}}$ and $\hat{\mathbf{V}}$ are more crucial since the deviations in $\mathbf{\Sigma}$ can be compensated with an one-tap equalizer at the receiver. These errors introduce interference between the transmitted parallel streams that can be called *inter-beam* or *inter-pattern interference* at both transmitter and receiver. Let's consider the virtual matrix representation ([2],[12]) of $\mathbf{H}_{bs}$ assuming an L-point uniform sampling $(L = 2\pi/\Delta\varphi)$ in the azimuth plane. The basis patterns at the virtual directions are given by the $(L \times M)$ matrices $\widetilde{\mathbf{B}}_{T/R}$. The virtual channel matrix [12] is defined by the following equation:

$$\mathbf{H}_{bs} = \widetilde{\mathbf{B}}_R^H \widetilde{\mathbf{H}}_b \, \widetilde{\mathbf{B}}_T \tag{24}$$

If $\boldsymbol{\Xi}_u, \boldsymbol{\Xi}_v$ are the estimation errors of $\hat{\mathbf{U}}$ and $\hat{\mathbf{V}}$ respectively then after pattern adaptation using the channel estimate, the channel matrix is given by:

$$\hat{\mathbf{U}}^H \mathbf{H}_{bs} \widetilde{\mathbf{V}} = (\mathbf{U} + \boldsymbol{\Xi}_u)^H \widetilde{\mathbf{B}}_R^H \mathbf{H}_b \widetilde{\mathbf{B}}_T (\mathbf{V} + \boldsymbol{\Xi}_v) = \boldsymbol{\Sigma} +$$
$$+ \left(\boldsymbol{\Xi}_u^H \widetilde{\mathbf{B}}_R^H \right) \mathbf{H}_b \widetilde{\mathbf{B}}_T \mathbf{V} + \mathbf{U} \widetilde{\mathbf{B}}_R^H \mathbf{H}_b \left(\widetilde{\mathbf{B}}_T \boldsymbol{\Xi}_v \right) + \boldsymbol{\Xi}_u^H \mathbf{H}_{bs} \boldsymbol{\Xi}_v \tag{25}$$

For high SNR, the last term with the product of the error matrices can be omitted as insignificant. The first error term expresses the inter-pattern interference due to the Rx adaptation while the second term due to Tx pattern adaptation. Therefore, the *interfering patterns* can be defined. The interfering patterns are the estimation error terms from the singular vector matrices expressed in the angular domain. If the adapted channel-aware basis patterns beamform the signal towards different angles, physical compensation of the interfering patterns is offered, since the interfering energy will be significantly reduced. This means that in channels with big and discrete scatterer clusters, the effects of the estimation error will be smaller due to the angle separation.

In Fig. 5a an example of the estimated reconfiguration patterns is presented compared with the ideal patterns that were calculated assuming perfect channel knowledge for a 'B2' channel of the 3-element ESPAR system. This channel has 2 aDoF and thus provides 2 channel-adapted orthogonal patterns. Estimation was performed with 13 dB mean MISO SNR. The SVD reallocates the power resources in favour of the first pattern in terms of SNR. Thus, the orthogonal path supported by the first pattern provides SNR = 16 dB while the second pattern provides SNR = 5 dB. Fig.5a shows that the first pattern is estimated perfectly. This is also true when mean MISO SNR is reduced to 7 dB, where the path supported from the first pattern achieves SNR=12dB. However, the second pattern produces high error values in low SNR, which is not quite clear by the investigation of the absolute values of the patterns in Fig.5a. Fig.5c shows that the interfering pattern is actually more powerful than the basis pattern itself due to significant random phase shifts. This is caused as a mathematical reaction of the SVD in the presence of powerful noise. Nevertheless the total squared error of $\mathbf{H}_{bs}$ will remain relatively low because

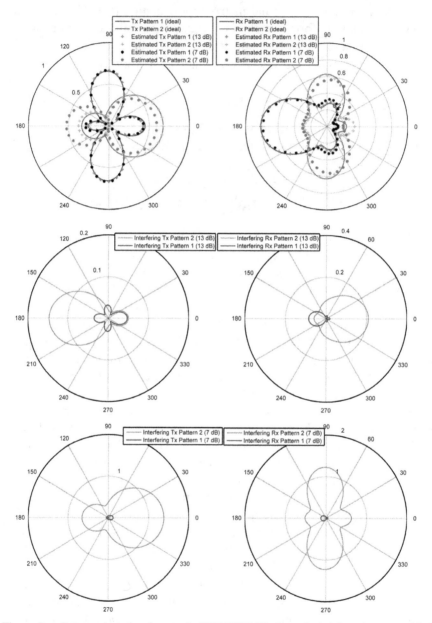

Figure 5   : Pattern adaptation for a typical WINNER B2 channel: a) Adapted patterns, ideal and via estimation (13, 7 dB SNR) b) Interfering patterns for the example (13 dB) c) Interfering patterns for the example (7 dB), [8]

the term $\Xi_u^H \mathbf{H}_{bs} \Xi_v$ of (25) has now significant power and opposes the error increase. In Fig.5b, a typical case is presented where the second interfering pattern does not aim towards the same direction with the first pattern and therefore, the effect of the estimation error that occurs is practically zero.

The result in (25) can be used to calculate the SNR degradation due to the channel estimation error in a BS-MIMO system with adaptive pattern reconfiguration. More particularly, if $\mathbf{E}$ is the $M \times M$ matrix that contains the inter-beam interference terms, given by:

$$\mathbf{E} = \left(\Xi_u^H \widetilde{\mathbf{B}}_R^H \right) \mathbf{H}_b \widetilde{\mathbf{B}}_T V + \mathbf{U} \widetilde{\mathbf{B}}_R^H \mathbf{H}_b \left(\widetilde{\mathbf{B}}_T \Xi_v \right) + \Xi_u^H \mathbf{H}_{bs} \Xi_v \qquad (26)$$

then the total BS-MIMO inter-pattern interference can be calculated by:

$$\varepsilon_{\mathrm{MIMO}} = \|\mathbf{E}\|_F^2 - (\mathrm{Tr}\,[\mathbf{E}])^2 \qquad (27)$$

and the inter-pattern interference caused at the $i-$th receiving beam (MISO interference) is given by:

$$\varepsilon_{MISO}^i = \sum_{k=1, k \neq i}^{M} \left| [\mathbf{E}]_{i,k} \right|^2 \qquad (28)$$

It is noted that the results in (27) and (28) do not include self-inflicted distortion that is expressed by the diagonal elements of $\mathbf{E}$. This error, if detected, can be removed by the system equalizer.

4 Evaluation of BS-MIMO through Simulation

In order to evaluate the performance of a BS-MIMO system that includes pattern adaptation and realistic channel estimation algorithms through simulation, two equivalent pairs of systems were assumed: a conventional MIMO system with 3-element or a 5-element ULA vs. the BS-MIMO system with a 3-element or a 5-element ESPAR antenna. The inter-element distance was set to $d = \lambda/16$ for all systems. Moreover, both systems use SVD for precoding or pattern reconfiguration as well as a similar channel estimation techniques (45 and 100 training samples for 3-element and 5-element systems respectively) as described in [11] and Sec.3. The systems use a simple coding rule that ensures BER $< 10^{-4}$ in order to achieve a certain level of adaptivity and exploit in a simple way the radio resources. Convolutional coding was applied using the default MATLAB

Trellis structure poly2trellis (7,[171 133],171). The applied automatic modulation and coding rule is presented in Tab. 1. Each time an estimated singular value indicates a channel with SNR<5.85 dB, the corresponding pattern is rejected and DoF/aDoF are reduced. It is noted that optimization of the adaptive coding and/or modulation rule was not an objective of this study.

A set of 3,000 flat 'B2' channels was applied to both systems. Moreover, in [19] it is stated that beamspace reception is performed in an M times oversampling rate, thereby noise power is increased by M at the BS-MIMO receiver.

Noise power was defined from the target mean BS-MIMO SNR. Every WINNER beamspace channel is normalized with its Forbenius norm: $\bar{\mathbf{H}}_{bs} = \mathbf{H}_{bs}/\|\mathbf{H}_{bs}\|_F$. In order to compare equivalent systems, the respective conventional MIMO matrix (that is produced by the same clusters of scatterers [13]) is also normalized by the same factor multiplied by $\sqrt{\bar{M}}$ ($\bar{\mathbf{H}} = \sqrt{\bar{M}}\mathbf{H}/\|\mathbf{H}_{bs}\|_F$), where $\bar{M}$ are the identified aDoF of the beamspace system. This means that, since $\bar{M}$ patterns are transmitted when adaptive pattern reconfiguration is used, the oversampling rate at the BS-MIMO receiver will also be reduced from M to $\bar{M}$. Consequently, the relative noise increase compared to the conventional system will also be reduced.

The results of this section are provided vs. the mean SNR in the beamspace per MISO channel. If P is the available transmitted power, then the mean BS-MISO SNR is defined as:

$$\text{SNR}_{bs,\text{MISO}} = \frac{P}{M\sigma_n^2} \tag{29}$$

The SNR value of (29) is used as a reference and as the x-axis in the provided figures of the section. Based on the above analysis, the corresponding SNR for the conventional MIMO system is given by:

$$\text{SNR}_{conv,\text{MISO}} = \frac{\bar{M} \times \text{SNR}_{bs,\text{MISO}} \|\mathbf{H}\|_F^2}{\|\mathbf{H}_{bs}\|_F^2} \tag{30}$$

Table 1   The Automatic Modulation and Coding rule.

SNR Range	Modulation and Coding
$SNR < 5.85\,dB$	Stream unused
$5.85\,dB < SNR < 8.5\,dB$	1/2-Convolutional Coding and QPSK
$8.5\,dB < SNR < 11.5\,dB$	3/4-Convolutional Coding and QPSK
$11.5\,dB < SNR < 15.2\,dB$	1/2-Convolutional Coding and 16 QAM
$15.2\,dB < SNR < 20.5\,dB$	3/4-Convolutional Coding and 16 QAM
$SNR > 20.5\,dB$	3/4-Convolutional Coding and 64 QAM

During this study, a method to bypass the drawback of the BS-MIMO noise level was identified through proper pulse shaping and filtering. However, since it is work-in-progress, the current evaluation includes the $\bar{M}$-times noise power increase in BS-MIMO.

The instantaneous MISO SNR for each channel in SVD-based systems varies significantly from the value in (29) depending on: a) the singular value of the specific MISO channel and b) the power allocation rule (e.g. waterfilling) and the power adaptation that ensures that Tx power will remain equal to P when less than M DoF/aDoF are available. An iterative rule is used to determine the available channels and the corresponding SNR. If the SNR $<$ SNR$_{\text{thres}}$ for the channel that corresponds to the smallest singular value σ_{min}, the channel is rejected and the power for the rest of the channels is adjusted to: SNR $\rightarrow$ $p_{\sigma_{min}} M/(M-i) \times$ SNR ($i = 1$). Then, the smallest remaining singular value is checked and $i \rightarrow i+1$, until all values are greater than SNR$_{\text{thres}}$(=5.85 dB for these scenarios). The algorithm is described in Algorithm 1.

A noise power estimation stage is also incorporated in the receiver allowing the estimation of SNR degradation due to the channel estimation error. The results in terms of spectral efficiency are presented in Fig. 6 for the 5-element

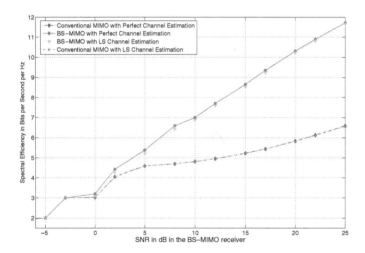

Figure 6 : Performance comparison between 5 element BS-MIMO and conventional MIMO systems in terms of Spectral Efficiency (net bit rate in the Physical Layer per Hz). The actual net bit rate is given by the value on the y-axis $\times$ system bandwidth (Perfect channel knowledge and LS estimation)

Algorithm 1: DoF/aDoF and assignment of coding rule

Calculate SVD from Channel Estimate
Perform Power Allocation with Waterfilling and calculate p_n
Determine SNR for every MISO Channel, $q \rightarrow 0$, $i \rightarrow 0$
While $q = 0$ & $i < M$ **DO**
    If $SNR > SNR_{thres}$ AND $p_\sigma > 0$ for all $M - i$ channels **Then**
        $aDoF = M - i$, Set coding rule for each channel
        $q \rightarrow 1$
    else
        Drop the path σ_{min} of the smallest singular value from the $M - i$ channels
        $i \rightarrow i + 1$
        increase $SNR \rightarrow p_{\sigma_{min}} M/(M - i) \times SNR$
    endif
    endwhile

---

systems and in Fig. 7 for the 3-element systems. BS-MIMO outperforms the conventional system, especially in high SNR. It is also noted that regardless of the channel or the noise power the conventional system rarely exceeds

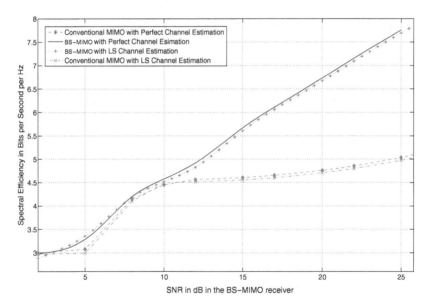

Figure 7   : Performance comparison between 3 element BS-MIMO and conventional MIMO systems in terms of Spectral Efficiency (net bit rate in the Physical Layer per Hz). The actual net bit rate is given by the value on the y-axis ×system bandwidth (Perfect channel knowledge and LS estimation)

the 2 DoF. Moreover, in many cases the 3-element conventional system degenerates to SISO. BS-MIMO on the other hand manages to achieve parallel transmission in the majority of the cases for high SNR. As noise power increases, BS-MIMO converges to the conventional MIMO behaviour maintaining a small advantage. At low SNR, both BS-MIMO and conventional MIMO turn into SISO and are set to operate with the same rules, therefore their performance is identical. Some steps or plateaus that can be seen in the figures are the result of the used adaptive modulation and coding rule. Another observation is that estimation appears to work less efficiently in the BS-MIMO case since the deviation of bits per symbol from the respective curve with perfect channel state information is higher. However, this is not accurate. It is reminded that noise is increased by M in the training period of the BS-MIMO system (regardless of the $\bar{M}$ aDoF, since estimation is performed for all M MISO channels) and therefore the difference in the performance of the estimator is due to the increased mean squared error caused by the inherent SNR degradation. In fact the performance of the estimator seems to be adequate despite of the increased by 7 dB noise level for the 5-element system and 4.77 dB for the 3-element system. Results for the 3-element BS-MIMO are also provided in [8]. Nevertheless, it must be noted that: a) in [8] only QPSK modulation is used, therefore the spectral efficiency remains low, while in Fig. 7, a powerful channel may use up to 64-QAM, b) in [8] no power allocation algorithm (waterfilling) is used and c) in [8] E_b/N_o is used as the x-axis which compared with the MISO SNR leads to a shift of values by 3dB for QPSK modulation.

It is also clear that if the BS-MIMO system was not subjected to the increased noise level, the achieved bit rate would be even grater. In low SNR, where aDoF $\rightarrow$ 1, the noise level remains the same for both systems since no oversampling is required for the BS-MIMO receiver operation. However, in high and mainly in medium SNR ranges, lower noise would mean more aDoF, less coding redundancy and better estimation results. Noise reduction can be achieved with proper pulse shaping of the transmitted pulses and with the use of a filtered downsampling procedure in the receiver. This architecture will limit the noise increase factor to a small value (e.g. 1.5 depending on the filter transition bandwidth).

It is also noted that when perfect channel knowledge is assumed, the BER target (10^{-4}) was achieved for all cases. On the other hand, channel estimation error and the consequent SNR degradation resulted in slightly increased BER level, which means that in a limited number of cases the BER threshold for both BS and conventional MIMO system was exceeded.

Finally given the fact that the used adaptive modulation and coding scheme is simple and not optimal, it is expected due to the existing theoretical capacity results [6] that the BS-MIMO bit rate advantage can become even grater.

5 Conclusions

This paper dealt with the subject of BS-MIMO channel estimation. Analysis of MMSE and LS estimators was performed through simulation. It was concluded that the estimation procedures are quite similar with conventional MIMO algorithms. The concept of interfering patterns was introduced and its physical meaning was presented. Finally, an attempt to evaluate BS-MIMO through link level simulation was made. BS-MIMO transceivers with adaptive pattern reconfiguration and practical, realistic reception algorithms that operate in WINNER 'B2' channels were assumed. It was confirmed that BS-MIMO systems outperform conventional MIMO for small interelement distances despite the increase of noise power due to oversampling. Prevention of the noise level increase is a subject for future work.

Acknowledgment

This paper is an extension of the work entitled "Channel Estimation and Link Level Evaluation of Adaptive Beamspace MIMO Systems" [8] that received the Best Paper Award in the Wireless VITAE conference of the Global Wireless Summit 2013. The research is co-financed by the European Union (European Social Fund-ESF) and Greek national funds through the Operational Program "Education and Lifelong Learning" of the National Strategic Reference Framework (NSRF) - Research funding Program: THALES Invensting in knowledge society through the European Social Fund, MIS379489.

References

[1] O. N. Alrabadi, J. Perruisseau-Carrier and A. Kalis. MIMO Transmission Using a Single RF Source: Theory and Antenna Design. In *IEEE Transactions on Antennas and Propagation*, vol.60, no.2, pp.654–664, 2012.

[2] A. Kalis, A. Kanatas and C. Papadias. A Novel Approach to MIMO Transmission Using a Single RF Front End. In *IEEE Journal on Selected Areas in Communications*, vol. 26, no 6, pp.972–980, August 2008.

[3] M. Wennstrom and T. Svantesson. An antenna solution for MIMO channels: the switched parasitic antenna. In *12th IEEE International Symposium on Personal, Indoor and Mobile Radio Communications*, pp.159–163, vol.1, 2001.

[4] T. Ohira and K. Gyoda. Electronically steerable passive array radiator antennas for low-cost analog adaptive beamforming. In *Proceedings of IEEE International Conference on Phased Array Systems and Technology*, pp.101–104, 2000.

[5] V. Barousis and A. G. Kanatas. Aerial degrees of freedom of parasitic arrays for single RF front-end MIMO transceivers. In *Progress In Electromagnetics Research B*, vol. 35, pp.287–306, 2011.

[6] P. N. Vasileiou, K. Maliatsos, E. D. Thomatos and A. G. Kanatas. Reconfigurable Orthonormal Basis Patterns Using ESPAR Antennas. In *IEEE Antennas and Wireless Propagation Letters*, vol.12, pp.448–451, 2013.

[7] V. Barousis, A. G. Kanatas, A. Kalis and C. Papadias. A Stochastic Beamforming Algorithm for ESPAR Antennas. In *IEEE Antennas and Wireless Propagation Letters*, vol.7, pp.745–748, 2008.

[8] K. Maliatsos, P. N. Vasileiou and A. G. Kanatas. Channel Estimation and Link Level Evaluation of Adaptive Beamspace MIMO Systems. In *Proceedings of Wireless VITAE, Global Wireless Summit*, 2013.

[9] A. Paulraj, R. Nabar and D. Gore. *Introduction to Space-Time Wireless Communications*. Cambridge University Press NY-USA, 2008.

[10] E. Thomatos, P. Vasileiou, and A. Kanatas. ESPAR loads calculation for achieving desired radiated patterns with a genetic algorithm. *The 7th European Conference on Antennas and Propagation, EuCAP*, 2013.

[11] O. Weikert and U. Zolzer. Efficient MIMO Channel Estimation With Optimal Training Sequences. In *in Proceedings of 1st Workshop on Commercial MIMO-Components and Systems (CMCS 2007)*, 2007.

[12] A. M. Sayeed. Reconfigurable Orthonormal Basis Patterns Using ESPAR Antennas. In *IEEE Transactions on Signal Processing*, vol.50, no.10, pp.2563–2579, 2002.

[13] K. Maliatsos, A. G. Kanatas. Modifications of the IST WINNER channel model for beamspace processing and parasitic arrays. In *Proceedings of the 7th European Conference on Antennas and Propagation, EUCAP 2013*, 2013.

[14] L. Hentila, P. Kyosti, M. Kaske, M. Narandzic and M. Alatossava. MATLAB implementation of the WINNER Phase II Channel Model ver1.1. In *https://www.ist-winner.org/phase_2_model.htm*, 2007.

[15] T. Ohira and K. Iigusa. Electronically steerable parasitic array radiator antenna. In *Electronics and Communications in Japan (Part II: Electronics)*, pp.25–45, Wiley Subscription Services, 2004.

[16] V. I. Barousis, A. G. Kanatas and A. Kalis. Beamspace-Domain Analysis of Single-RF Front-End MIMO Systems. In *IEEE Transactions on Vehicular Technology*, vol.60, no.3, pp.1195–1199, 2011.

[17] J. Pang, J. Li, L. Zhao and Z. Lu. Optimal Training Sequences for MIMO Channel Estimation with Spatial Correlation. In *Proceedings of the 66th IEEE Vehicular Technology Conference, VTC-2007 Fall*, pp.651–655, 2007.

[18] X. Ma and L. Yang and G. B. Giannakis. Optimal training for MIMO frequency-selective fading channels. In *IEEE Transactions on Wireless Communications*, pp.453–466, 2005.

[19] R. Bains and R. R. Muller. Using Parasitic Elements for Implementing the Rotating Antenna for MIMO Receivers. In *IEEE Transactions on Wireless Communications*, vol.7, no.11, pp.4522–4533, 2008.

[20] P. Kyosti, J. Meinila, L. HentilaBains, et all. *IST-4-027756 WINNER II D1.1.2 v.1.1: WINNER II channel models*, 2007

Biographies

Konstantinos Maliatsos received his Diploma in Electrical and Computer Engineering from the National Technical University of Athens, Greece (NTUA) in 2003. He joined the NTUA Mobile Radio-Communications Laboratory (MRCL) working as an assistant researcher in various industry and research oriented projects. In 2005 he earned his Master's degree in Business Administration from the postgraduate studies program "Techno-Economic Systems" organized by NTUA, National Kapodistrian University of Athens and University of Piraeus (UniPi), while he started working on his PhD entitled "Transmission techniques, Design and Analysis of Software Defined Radio - Cognitive Radio Adaptive wireless transceivers". He received his PhD degree in 2011. As an assistant researcher he worked for 3 years on projects for the Greek General Secretariat for Research and Technology on Cognitive Radios and also participated in MRCL European projects. Currently he is working on his post-doctoral research on advanced MIMO techniques in the Department of Digital Systems of UniPi. He also continuous to cooperate with MRCL as a researcher for FP7 European projects. He has published more than 25 papers in International Journals and Conference proceedings. His research interests include Software Radio, Cognitive Radio and Dynamic Spectrum Access, Multicarrier Modulations, MIMO systems (conventional, beamspace, massive and distributed), Filter Bank theory, Synchronization for broadband wireless, Channel Modeling, Detection and Estimation theory.

Panagiotis N. Vasileiou received his M.Sc. Degree in 2012 in digital communications and networks with honors and the B.Sc. Degree in 2010, from the Department of Digital Systems in the University of Piraeus. He has been awarded with scholarships for the first place in the academic records in three several periods during the B.Sc. program and one scholarship during the M.Sc. Since November 2013, he is a Phd student at the Telecommunication System Laboratory, Department of Digital Systems, University of Piraeus, where he involved in various research projects with published work in various communities.

Athanasios G. Kanatas received the Diploma in Electrical Engineering from the National Technical University of Athens (NTUA), Greece, in 1991, the M.Sc. degree in Satellite Communication Engineering from the University of Surrey, Surrey, UK in 1992, and the Ph.D. degree in Mobile Satellite Communications from NTUA, in February 1997. From 1993 to 1994 he was with National Documentation Center of National Research Institute. In 1995 he joined SPACETEC Ltd. In 1996 he joined the Mobile Radio-Communications Laboratory as a research associate. From 1999 to 2002 he was with the Institute of Communication & Computer Systems. In 2000 he became a member of the Board of Directors of OTESAT S.A. In 2002 he joined the University of Piraeus where he is a Professor in the Department of Digital Systems. From 2008 to 2009 he has served as Member of the Senate of University of Piraeus. From 2007 to 2009 he served as Greek Delegate to the Mirror Group of the Integral Satcom Initiative. His current research interests include the development of new digital techniques for wireless and satellite communications systems, channel characterization, simulation, and modeling for future mobile, mobile satellite and wireless communication systems, antenna selection and RF preprocessing techniques, new transmission schemes for MIMO systems, and energy efficient techniques for Wireless Sensor Networks. He has published more than 120 papers in international Journals and conference proceedings. He has been a Senior Member of IEEE since 2002. From 1999 to 2009 he chaired the IEEE ComSoc Chapter of the Greek Section. This year he elected Dean of the School of Information & Communication Technologies of the University of Piraeus.

New Efficient Timing and Frequency Error Estimation in OFDM

A. P. Rathkanthiwar[1] and A. S. Gandhi[2]

[1]*Department of Electronics Engineering, Priyadarshini College of Engineering, Nagpur, MS, India, anagharathkanthiwar@yahoo.co.in*
[2]*Professor, Department of Electronics, Vesvesvarya National Institute of Technology, Nagpur, MS, India, abhay4083@yahoo.co.in*

Received 24 August 2013; Accepted 27 November 2013;
Publication 23 January 2014

Abstract

Orthogonal frequency division multiplexing (OFDM) technique is being applied extensively to high data rate digital transmission as it is a bandwidth efficient modulation scheme. In OFDM system, intersymbol interference (ISI) and intercarrier interference (ICI) occur due to synchronization errors. Fast, simple and robust synchronization algorithms are necessary for OFDM systems. In OFDM downlink transmission, every terminal perform synchronization by exploiting reference symbols called training symbols of received frame. In this paper we are presenting very simple method of synchronization which uses training symbol to determine timing and frequency synchronization error. Proposed synchronization techniques result in better performance with respect to all the parameters despite being simple.

Keywords: Synchronization, orthogonal frequency division multiplexing (OFDM), timing estimation, CFO, CFO estimation.

1 Introduction

Orthogonal frequency division multiplexing (OFDM) is a bandwidth efficient modulation scheme for high speed data communication in frequency-selective multi-path fading channels [1]. The mitigation of frequency selectivity in multi-path fading channels is possible with OFDM, because the frequency

Journal of Cyber Security, Vol. 2 No. 3 & 4 , 291–306.
doi: 10.13052/jcsm2245-1439.235

selective fading channel is transformed into multiple flat fading channels, one for each sub-carrier. The orthogonality among sub-carriers makes a more compact signal bandwidth for a given data rate. Furthermore, it provides larger flexibility by allowing independent selection of the modulation parameters (like the constellation size and coding scheme) over each sub-carrier. OFDM Modulation can be realized with Inverse Fast Fourier Transform (IFFT). Due to all these favorable features, many digital transmission systems have adopted OFDM as the modulation technique such as digital video broadcasting terrestrial TV (DVB-T), digital audio broadcasting (DAB), terrestrial integrated services digital broadcasting (ISDB-T), digital subscriber line (xDSL) etc. Now it is being used in packet based systems like multimedia mobile access communications (MMAC), and the fixed wireless access (FWA) system in IEEE 802.16.3 standard [2]. It has become fundamental technology in the future 3GPP LTE and 4G-multimedia mobile communication systems.

However, the OFDM transmission is sensitive to receiver synchronization imperfections. The symbol timing synchronization error cause intersymbol interference (ISI) and the frequency synchronization error is one of the reasons for intercarrier interference (ICI). In OFDM, frequency synchronization errors are actually Carrier Frequency Offsets (CFO) and are generally caused by unmatched local oscillators at the two ends of the communication links, Doppler shifts or phase noise introduced by nonlinear channel[3]. Thus, synchronization is a crucial issue in an OFDM receiver design. Task of timing synchronization of an orthogonal frequency division multiplexing (OFDM) receiver, requires alignment of the discrete Fourier transform (DFT) segments with OFDM symbol boundaries [4]. Timing alignment errors may occur in cases where the DFT aperture contains part of the guard interval that has been distorted by intersymbol interference (ISI). Synchronization techniques for OFDM systems can be classified as either blind or data-aided. Blind approaches exploit the inherent redundancy in the OFDM signal structure for example, cyclic prefix (CP). Even though blind techniques have the advantage of not requiring extra overhead, their performance usually degrades when the noise level is high or the channel distortion is severe. Data-aided techniques offer the advantage of superior performance in low SNR applications at the expense of reduced spectral efficiency[4]. These techniques make use of reference symbols embedded into the transmitted signal. One way of embedding a reference symbol into a transmitted signal is to prefix it to the beginning of the time-domain waveform in the form of a preamble. Reliable synchronization is one of the key factors determining the transmission performance in communication channels. Various schemes for time and frequency estimation for imperfect communication links have been explained in the literature.

System Model

In OFDM systems, the data is modulated in blocks by means of a Fast Fourier transform (FFT). Data stream is mapped into N complex symbols in frequency domain. These N complex symbols are modulated on N subcarriers by using N-point inverse fast Fourier transform (IFFT) and the time domain samples are computed using the well-known IFFT formula

$$x(n) = \frac{1}{\sqrt{N}} \sum_{k=0}^{N-1} d_k(n)\, e^{\frac{j2\pi kn}{N}} \quad n = 0, 1, ...N - 1 \tag{1}$$

Received signal can be modeled as

$$r(n) = x(n - \theta)e^{j2\pi n\epsilon} + w(n) \tag{2}$$

Here θ represents the unknown integer-valued time offset, ε is the carrier frequency offset (CFO), w(n) is complex additive white Gaussian noise and x(n) is transmitted signal.

Timing Error Estimation

Timing error or timing offset θ occurs because of channel delay and multipath dispersion. Due to this timing error the receiver's time-domain FFT window spans samples from two consecutive OFDM symbols. This results in inter-OFDM symbol interference leading to BER degradation [6]. The requirements on the time offset estimator are determined by the difference in length between the cyclic prefix (CP) and the channel impulse response (CIR) [5]. This difference is the part of the cyclic prefix that is not affected by the previous symbol due to the channel dispersion, as shown in Figure 1. As long as a symbol time offset estimate does not exceed this difference, the orthogonality of the subcarriers is preserved, and a time offset within this interval only results in a phase rotation of the subcarrier constellations. The closer the time offset estimate is to the true offset, the shorter the cyclic prefix needs to be, reducing the overhead in the system.

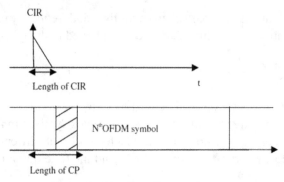

Figure 1   The time offset requirements

Effect of Timing Error

The effect of uncompensated timing error on system performance can be described by figure 2. As shown in figure 2 the tail of each received block extends over the first L - 1 samples of the successive block as a consequence of multipath dispersion. If cyclic prefix is greater than CIR duration (0 to L-1) then there is certain interval which is not affected by previous block. As long as the DFT window starts anywhere in this interval, no Inter Symbol Interference (ISI) is present at the FFT output. This situation occurs whenever the timing error θ belongs to interval $-Ng + L - 1 \leq \theta \leq 0$ and only results in a cyclic shift of the received OFDM block [7]. Thus, recalling the time-shift property of the Fourier transform and assuming perfect frequency synchronization the DFT output over the nth subcarrier takes the form

$$R(k) = e^{\frac{j2\pi k\theta}{N}} H(k)d(k) + w(k) \tag{3}$$

Here

$$H(k) = \sum_{l=0}^{L-1} h(l)\, e^{-j2\pi kl/N} \tag{4}$$

Where $h(l)$ is complex channel impulse response, d corresponds to OFDM symbol.

The equation of R (k) indicates that timing error θ appears as a linear phase across subcarriers. On the other hand, if the timing error is outside the interval $-Ng + L - 1 \leq \theta \leq 0$, samples at the FFT input will be contributed by two adjacent OFDM blocks causing Inter Symbol Interference (ISI). In addition to ISI, this results in a loss of orthogonality among subcarriers which in turn,

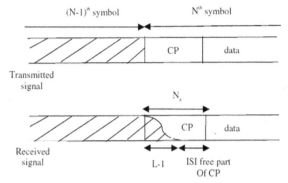

Figure 2 Partial overlapping between received blocks due to multipath dispersion

generates Inter Carrier Interference (ICI). In this case, the kth DFT output is given by

$$R(k) = e^{\frac{j2\pi k\theta}{N}} \alpha(\theta)H(k)d(k) + I(k,\theta) + w(k) \qquad (5)$$

Where I (k,θ) is the introduced ISI. Due to the introduced ISI and the phase rotation, there is slight magnitude attenuation α (θ) in the signal.

Timing Error Estimation

In the process of synchronization in OFDM systems, errors or offsets are estimated and then corrected using the estimated values. In order to perform the FFT demodulation correctly, the symbol timing synchronization must be done to determine the starting point (i.e. FFT window) of the OFDM symbol. It is important to solve symbol timing synchronization problem first during the design process of an OFDM receiver. The conventional algorithms for the symbol timing estimation in time domain are MLE (Maximum Likelihood Estimation) utilizing the cyclic prefix of the OFDM symbols. The most representative algorithm was proposed by J. J. Van de Beek [2]. However, good performance achieved only under the AWGN channel. When the channel condition becomes severely degraded, data is badly contaminated by ISI. T. M. Schmidl and Cox (S and C) introduced a new method making use of the training symbol called reference block which is placed in front of data symbols in the frame [7]. Training symbols or reference blocks has repetitive structure in the time domain and are exploited for error estimation purpose at the receiver. In this case, timing estimator can be designed by searching for the peak of the correlation among repetitive parts called as timing metric. Training symbol used by Schmidt and Cox (S & C) composed of two identical halves of length N/2. It has better performance compared to that proposed by J. J. Van de Beek. Unfortunately, timing metric of this algorithm exhibits a large plateau that may greatly reduce the estimation accuracy[6]. To reduce uncertainty due to timing metric plateau, Shi and Serpedin (S & S) used a training block composed of four repetitive parts [+ B + B – B + B] with a sign inversion in the third segment [8]. Here B is PN sequence of length N/4. More accurate timing estimate is obtained. Jung Ju Kim, Jungho Noh, and Kyung Hi Chang presented a scheme in their paper [9] which uses a training symbol composed of four parts [B B* B B*]. Here * represents conjugate of quantity. This method reduces plateau and it can achieve more accurate timing offset estimation. Since CFO is usually unknown at this stage, it is desirable that the timing recovery scheme be robust against possibly large frequency offsets. The first step of the timing estimation is the detection of a new frame in the received data stream. For this purpose, timing metric is continuously monitored and the start of a frame is declared whenever it overcomes a given threshold or when a peak is observed in the timing metric

Proposed Timing Error Estimator

The training symbol used for proposed algorithm is similar to the first training symbol used in [7] and has two identical halves each of length N/2 in time domain. This training symbol is used as a preamble. The Timothy M. Schmidl and Donald C. Cox (S & C) method [7] uses autocorrelation function (ACF) to compute timing metric. The formula is

$$(P)_{ACF}(n) = \sum_{k=0}^{\frac{N}{2}-1} r(n+k+N/2)r^*(n+k) \tag{6}$$

Note that n is a time index corresponding to the first sample in a window of N samples. This window slides along in time as the receiver searches for the first training symbol. For this purpose, timing metric is continuously monitored and the start of a frame is declared whenever it overcomes a given threshold. Unfortunately the timing metric of the S & C algorithm exhibits a large plateau that may greatly reduce the estimation accuracy. This timing metric is shown in figure 3 (a). However cross correlation function (CCF) gives better timing metric. The corresponding equation is

$$P_{CCF}(n) = \sum_{k=0}^{N-1} r(n+k) \ s^*(k) \tag{7}$$

Timing metric of this CCF function produces three peaks as shown in figure3 (b). Major peak corresponds to full symbol match and two minor peaks corresponds to half symbol match. Start of a frame is declared and timing error is estimated whenever maximum is detected. However minor peaks may cause trouble in severe noise conditions in the estimation process. Drawbacks of both ACF & CCF timing metric is overcome by multiplying both functions over fixed interval. The resultant timing metric shows single and clear major peak as shown in figure 4. This results in improved estimation performance. Metric of this proposed algorithm is obtained by using

$$P(n) \ = \ P_{ACF} * P_{CCF} \tag{8}$$

Arrival of frame is declared whenever a peak is detected and timing error is estimated by

$$\widehat{\theta} = \arg(\max(P(n))) \tag{9}$$

Here timing metric shown in figures 3 and 4 are obtained in the presence of timing offset. Timing error is estimated using proposed algorithms and S and C algorithm. The parameters used for performance analysis of these proposed timing error estimators are mean error occurred in timing error estimation, standard deviation of timing error estimation, and probability of correct detection.

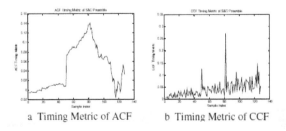

a Timing Metric of ACF   b Timing Metric of CCF

Figure 3 Timing Metric of ACF b Timing Metric of CCF

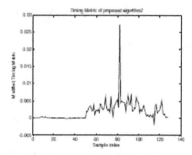

Figure 4   Timing Metric of proposed algorithm2

Frequency Offset Estimation

The FFT filter's frequency response extends over the whole frequency range and the very narrow spacing between sub-carriers all contribute to the sensitivity of OFDM systems to frequency synchronization errors[3]. Carrier frequency offset if present causes a phase rotation. This phase rotation effect is clearly seen through constellation diagram as shown in figure 5.

These constellation diagram are obtained for 4QAM in the presence of 30 dB SNR noise and subcarrier spacing is 312.5 KHz.

Using the fact that phase separation between the two consecutive peaks of CCF function is proportional to CFO. Carrier frequency offset hence can be estimated using [10]

$$\Delta f = \frac{\Delta \phi}{\pi \ T_u/2} \tag{10}$$

Here $\Delta \phi$ is phase difference between two peaks of CCF function which are separated by time duration of $T_u/2$.

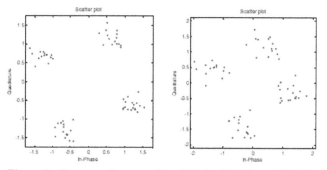

Figure 5   Phase rotation caused by CFO for 30 KHz and 50 KHz

Performance Evaluation

Simulation Parameters

Performance of the proposed synchronization technique is evaluated by MATLAB ® [1] simulation over the AWGN and frequency-selective fading channels typical of broadband wireless communication systems. Specifically, as a fading channel model, this paper uses ITU channel model [11] for Vehicular Test Environment of type A and details of this channel model is given in table 1 below.

The other major simulation parameters are listed in table 2. Simulation is performed to obtain Mean error of timing estimation, Standard deviation of timing estimation, Detection probability, Bit error ratio etc. for performance measurement of proposed algorithm. For simulation in MATLAB, known fixed timing offset is applied to the OFDM frame[9] and the channel used is AWGN and Rayleigh channel as defined in table 2.

Table 1 Rayleigh channel details

Tap	Channel A	
	Relative Delay (ns)	**Average Power (dB)**
1	0	0.0
2	310	−1.0
3	710	−9.0
4	1090	−10.0
5	1730	−15.0
6	2510	−20.0

Table 2 Simulation Parameters

FFT length N	64
Constellation mapping	4 & 16 QAM
Subcarrier Spacing in KHz	312.5
Useful OFDM Symbol Period	3.2e-6
Applied Timing Offset theta	18
Number of simulation runs	100
SNR range	0 to 20 dB
Number of Symbols	30

---

[1] MATLAB is a registered trademark of MathWorks Inc.

Symbol Timing Estimator Performance for AWGN Channel

Timing offset estimation mean error obtained with proposed algorithms as a function of SNR values for 4 QAM and 16 QAM is shown in figure 6. For comparison S and C algorithm have also been considered. For both algorithms, timing mean error decreases with increase in SNR value. However proposed algorithm becomes perfect estimator for SNR values of 10 dB and above in case of 4QAM.

Similarly figure 7 shows standard deviation graph of timing error estimation for 4QAM and 16QAM obtained using these algorithms. Detection probability graph are shown in figure 8. Detection probability increases with increase in SNR. Estimator corresponding to proposed algorithm achieves perfect synchronization for SNR value more than 10 dB. For 16QAM performance improvement of proposed algorithm is more with respect to detection probability. The graphs shown in figures 6,7, 8 are for AWGN channel.

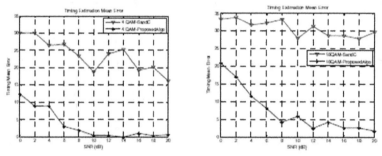

Figure 6   Timing Estimation Mean error for 4QAM and 16QAM

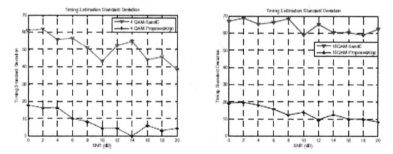

Figure 7   Timing Estimation Standard deviation for 4QAM and 16QAM

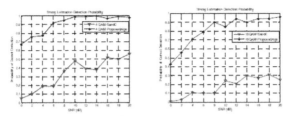

Figure 8   Detection Probability graph for 4QAM and 16QAM

Symbol Timing Estimator Performance for Rayleigh Channel

Results are also obtained for Rayleigh channel with parameters defined in table 1, using existing S and C algorithm and proposed algorithm. Resulting graphs for 4QAM and 16QAM are shown in figure 9, 10 respectively for mean error, standard deviation of timing error. From these figures it is noted that for Rayleigh environment also proposed algorithm give improved performance with respect to all parameters. Comparatively improvement in performance of proposed algorithm with respect to S and C algorithm is more in Rayleigh channel environment.

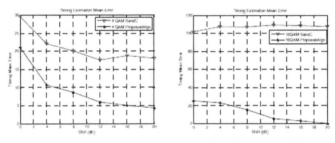

Figure 9   Timing Estimation Mean error for 4QAM and 16QAM

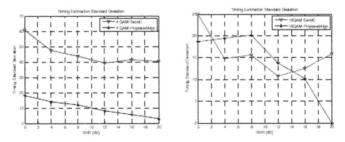

Figure 10   Timing Estimation Standard deviation for 4QAM and 16QAM

Overall BER Performance

To check overall system performance using proposed timing offset estimator, a BER graph is obtained taking AWGN channel for 4QAM. At receiver timing error is estimated using algorithm. Estimated value is used to correct timing error and after FFT and demodulation BER values are obtained. BER graph obtained is shown in figure 10.

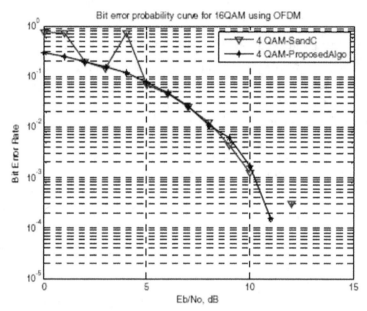

Figure 11   Bit Error Ratio graph for 4QAM, AWGN channel

As seen from figure 11, it is obvious that the proposed estimator significantly improves the BER performance compared to conventional S and C. It should be noted that, below 10 dB SNR, proposed estimators achieve more robustness than that of S and C.

Simulation for CFO Estimation

Known value of CFO is inserted in the symbol carriers. Simulation for CFO estimation is then done by assuming perfect timing synchronization. Value of CFO is estimated using equation 11 by first calculating phase between peaks of CCF function in the presence of noise. Result shown in table 3 are obtained using parameters of table 2 and 20 dB SNR. Same CCF metric of timing synchronization is used here to estimate CFO. Figure 12 indicates the graph

Table 3 Actual and estimated value of CFO

Inserted value of CFO in KHZ	Estimated value of CFO in KHZ
10	8.86
20	17.26
30	30.85
40	41.01
50	48.19
60	58.97
70	69.84
80	80.5
90	89.25
100	100.58

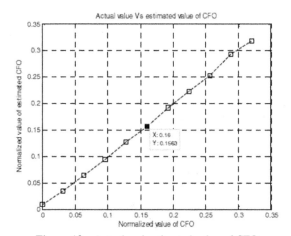

Figure 12   Actual and estimated value of CFO

of actual CFO versus estimated CFO. One data cursor in figure 12 is displaying both of these values. Values of CFO here are normalized with respect to sub carrier spacing.

Discussion and Conclusion

This paper examined and compared performance of proposed algorithm with existing S and C algorithm over AWGN and Rayleigh channel. This proposed timing synchronization algorithm is simple & the results are encouraging. Performance parameters like mean error, standard deviation of timing error estimation and detection probability are used. New algorithm shows superior performance to the existing S and C algorithm. Same training symbol and same function of timing estimation is used for CFO estimation. This reduces extra overhead and hence improves throughput of system.

References

[1] D. K. Kim, S. H. Do, H. B. Cho, H. J. Choi and K. B. Kim. A New Joint Algorithm Of Symbol Timing Recovery And Sampling Clock Adjustment For OFDM Systems, IEEE Transactions on Consumer Electronics, vol. 44, No.3, August 1998.

[2] J. J. van de Beek, M. Sandell, P. O. Borjesson. ML estimation of time and frequency offset in OFDM systems, Signal Processing, IEEE Transactions on [see also Acoustics, Speech, and Signal Processing, IEEE Transactions Volume 45, Issue 7, Page(s):1800–1805, July 1997.

[3] B. Ai, Member, IEEE, J.-h. Ge, Y. Wang, Member, IEEE, S. Y. Yang, and P. Liu. Decimal Frequency Offset Estimation in COFDM Wireless Communications, 154 IEEE Transactions on broadcasting, vol. 50, no. 2, June 2004.

[4] O. Ureten and Selc¸uk Tas¸cioglu. Autocorrelation Properties Of OFDM Timing Synchronization Waveforms Employing Pilot Subcarriers, Research Article EURASIP Journal on Wireless Communications and Networking.

[5] D. Landstrijml, S. Kate, W. J.-J. van de Beek. Per Odling' Per O. Borjesson'. Symbol time offset estimation in coherent OFDM systems" 1999 IEEE.

[6] M. K. Morelli, C.-C. J. Pun, M.-O. Synchronization Techniques for Orthogonal Frequency Division Multiple Access (OFDMA): A Tutorial Review, Proceedings of the IEEE, Volume: 95, Issue: 7, On page(s): 1394–1427, July 2007.

[7] T. M. Schmidl and D. C. Cox. Robust frequency and timing synchronization for OFDM, IEEE Trans. on Commun., vol. 45, no. 12, pp. 1613–1621, Dec. 1997.

[8] K. Shi and B. E. Serpedin. Coarse frame and carrier synchronization of OFDM systems: A new metric and comparison", IEEE Trans. Wireless Commun., vol. 3, no. 4, pp. 1271–1284, Jul. 2004.

[9] J. Kim, J. Noh, K. H. Chang. An efficient timing synchronization method for OFDMA system, Wireless Communications and Applied Computational Electromagnetics, 2005. IEEE/ACES International Conference on 3–7 April 2005 Page(s):1018–1021.

[10] H. Puska and H. Saarnisaari. Matched Filter Time and Frequency Synchronization Method for Ofdm Systems Using PN-Sequence Preambles, The 18th Annual IEEE International Symposium on Personal, Indoor and Mobile Radio Communications (PIMRC'07) University of Oulu, Centre for Wireless Communications (CWC).

[11] Channel Models : A Tutorial V1.0 by Raj Jain February 21, 2007.

Biographies

Abhay S. Gandhi did his BE (Electronic Engg) in 1989 from Visvesvaraya Regional College of Engineering (VRCE), Nagpur and ME (ECE) from Indian Institute of Science, Bangalore in 1991. After working for 3 years in industry and education, he joined VRCE (Now VNIT), Nagpur as Lecturer in July 1994. He has completed his PhD in August 2002. Currently, he is working as Professor at Department of Electronics Engineering, VNIT.

He has published 26 research papers in international conferences and journals. His research interests include signal processing, wireless digital communication, radio frequency (RF) circuits, and computer networks. E-mail: asgandhi@ece.vnit.ac.in

Anagha Rathkanthiwar did her BE (Electronic Engg) and M.Tech. (Electronics) in 1992 and 2003 respectively from RTMNU Nagpur (Nagpur University). She is pursuing her Ph.D. from Vesvesvaraya National Institute of Technology (VNIT), Nagpur. Currently, she is working as Assistant Professor

in Electronics Engineering Department, Priyadarshini College of Engineering, Nagpur, Maharashtra, India.

She has published 14 research papers in international conferences and journals. Her research interests include signal processing, wireless communication. She has reviewed several research papers of International and National conferences. Also she has chaired sessions at National Conferences. E-mail: anagharathkanthiwar@yahoo.co.in

Green Cooperative Web of Trust for Security in Cognitive Radio Networks

Vandana Milind Rohokale, Neeli Rashmi Prasad and Ramjee Prasad

Center for TeleInFrastruktur, Aalborg University, Aalborg, Denmark,
E-mail: vmr;np;prasad@es.aau.dk

Received 29 June 2013; Accepted 31 August 2013;
Publication 23 January 2014

Abstract

Spectrum is a scarce and very essential resource for the ever growing mobile communication applications. Radio networks (CRN) is the best evolved solution towards spectrum scarcity. Cooperative spectrum sensing is a well-known and proven mechanism in the CRNs. As compared to other traditional wireless networks, CRNs are more delicate and open to the wireless environments due to their heterogeneous nature. Therefore, the CRNs have more security threats than the conventional wireless networks. The spectrum sensing and sharing mechanisms are inherently vulnerable to the malicious behaviors in the wireless networks due to its openness. This paper proposes an energy efficient lightweight cryptographic Cooperative web of trust (CWoT) for the spectrum sensing in CRNs which is proved to be appropriate for the resource constrained wireless sensor networks (WSNs). Development of trust based authentication and authorization mechanism for the opportunistic large array (OLA) structured CRNs is the main objective of this paper. Received signal strength (RSS) values obtained can be utilized to avoid Primary user emulation attacks (PUEA) in CRNs.

Keywords: Cooperative Spectrum Sensing, Cognitive Radio Network (CRN), Cooperative Web of Trust (CWoT), Wireless Sensor Network (WSN), etc.

1 Introduction

Wireless communication and relative mobile computing applications is a boom in the telecom market but the available spectrum and its allocation is not appropriate to satisfy the highly increasing demands by mobile applications. Cognitive radio technology is a hopeful evolution for the solution towards scarce radio spectrum [1]. Cognitive radio entities continuously sense the spectrum holes which are utilized for the opportunistic communication. CRNs provide the spectrum reuse concept. Since the CRN evolves from the hybrid combination of many heterogeneous networks, it is much more prone to the wireless open media vulnerabilities [2]. Consumer premise equipment (CPE) which has the inbuilt cognition capability, continuously monitors the spectrum, senses the white spaces in the spectrum and occupies the spectrum according to the availability and it can vacate the occupied spectrum immediately after sensing the comeback of the licensed user.

CPE is a mobile equipment with cognition capabilities which can sense radio environment eg., spectrum white spaces, information about geographic location, available wireless or wired networks around and available services. It can also analyze and get information regarding the secondary user's needs and reconfigure itself by adjusting some specific parameters to make sure that rules and regulations of CRN are strictly followed. Whenever the CPE senses spectrum holes, CRN sends Request to send (RTS) kind of packets on the network to initiate the communication [3].

Cooperation is the vital characteristic of CRNs because the secondary nodes of CRN basically cooperate with each other for finding out the spectrum white spaces in the available spectrum for the successful and timely wireless communication. With cognitive environment, it is very much essential that the information bearing secondary nodes should exchange their data through multicast communication. Safety of the secondary user's communication data from intruder is a critical issue for CRNs. Because of these reasons, Group Security is necessary for secondary users of CRN [4]. The group based security with collaborative advantages is possible with the concept of Cooperative Web of Trust (CWoT).

CRN is a multi-user environment where multiple secondary and primary users are present in the system. Spectrum sensing for such multi-user case becomes more complicated wherein the sensing of spectrum holes and the interference estimation are the complex tasks. A collaborative effort by secondary users is the attractive solution. The research work in [5] proposes a new cooperative spectrum sensing mechanism for multi-user CRNs in which each user's contribution is weighted by taking into account the parameters like received power and path loss components.

The paper is organised as follows. Related works in the security of the CRN is discussed in section II. Section III explains the proposed system model for the green and cooperative web of trust mechanism for security in the cognitive radio networks. It explains in details how authentication, trust building and authorization are achieved in the CWoT system. Simulation results are depicted in section IV. This section shows energy efficiency of the proposed security system. Section V includes conclusions and future scope for this work.

2 Related Works

For wireless communication, a signal has to be transmitted through open media with a virtual connection. Since the CRNs are built with the numerous heterogeneous wired and wireless networks, the chances of the data being hijacked are more. Figure 1 below depicts the security threat taxonomy for CRN [6] wherein the possible security threats to CRN are mentioned.

O. Leon *et. al.* in [7], have studied various possible vulnerabilities to CRNs with classification of attacks and their impact. They have proposed security solutions to CRNs keeping in mind the FCC rules and regulations regarding primary user system and their services should remain intact irrespective of modifications in the secondary users of CRNs. In [8], cross layer (Physical plus MAC) attack strategies such as coordinated report false sensing data

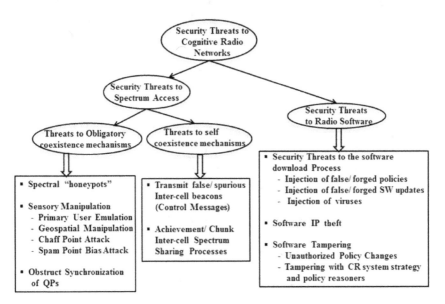

Figure 1 Security Threats Taxonomy for CRNs

attack and small back-off window attack are designed and a trust based cross layer defense framework is proposed. Due to the security provided by proposed defense mechanism, the damage percentage is shown to be reduced.

Trust based security system for community based CRNs is proposed in [9]. Here the trust value of a CR node is decided according to the history behavior of that node. Here the authors have designed trust based authentication for community based CR nodes. Paper [10] puts forth a trust based algorithm for CRNs which is based on location consciousness and estimated distance between the mobile users. Here the trust calculation is performed based on received power and trust metrics are decided by the combination of trustworthiness requirements and QoS of the radio links.

In the research work of [11], the authors have calculated trust depending on various communications attributes and it is compared with the threshold value of trust. Helena et. al. in [12] have presented a good combination of wireless physical layer security, private key cryptography and one way hash functions. They have proposed a security protocol for a centralized system where the authenticity is verified at the data fusion center which they claim as the robust mechanism against the location disclosure attacks. The research work in [13] proposes a trust methodology for secrecy in cooperative spectrum sensing (TM-SCSS) wherein the data fusion centre assigns and updates the trust value to each entity according to the sensing results. The secondary cooperating

Figure 2   Gain and Overheads in Cooperative Spectrum Sensing [14]

radio nodes are classified into categories like malicious node, pending node and trusted node based on their recent trust values updated according to the data fusion center.

Cooperative spectrum sensing is a dominant technique for the detection mechanism in the CRN. It makes use of cooperative spatial diversity to exploit benefits like energy efficiency, cooperation efficiency and wideband sensing capability. But the advantages come with certain overheads like security challenges due to heterogeneous nature of CRN, sensing time and delays, mobility management and channel impairments as depicted in Figure 2. The techniques used for spectrum sensing include Energy Detector based Sensing, Cyclostationary based sensing, Radio Identification based sensing and matched filtering. For energy detection based sensing, cooperation is the best suited technique since it results into appropriate received signal strength values.

2.1 Primary User Emulation Attack (PUEA)

An attacker imitates the characteristics of a primary signal transmitter and pretends as being primary user as shown in Figure 3. Proper identification mechanisms are very much essential for the prevention of the PUEA attacks in CRN. The problems associated with the PUEA attack are security related, trust related and also performance related. So, for the prevention of the critical threat like PUEA, some kind of strong security mechanism is vital [6].

Game theoretic cooperation approaches promise to provide proper incentives for the nodes cooperating to relay the information from sender to receiver. Mainly, three types of behaviours are observed in the wireless networks like

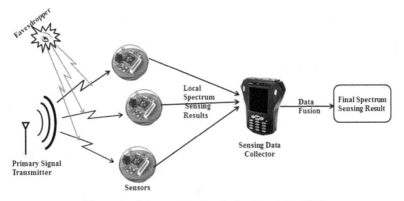

Figure 3 Primary User Emulation Attack in CRN

No Help (Egoistic Behaviour), Unidirectional Help (Supportive Behaviour) and Mutual Help (Cooperative Behaviour). For the construction of trust based security we have taken into consideration this incentive aspect in terms of increment in the trust level for good cooperation and trusted behavior. This paper proposes a lightweight CWoT for the prevention of primary user emulation attacks in the spectrum sensing technique for the heterogeneous cognitive radio networks. The facts with CWoT's considerably improved received signal strength (RSS) figures ensure the security against identity thefts of the primary users. With the CWoT mechanism, authentication and authorization techniques are proposed which are based on trust levels. The secondary user's cognitive radio equipment forms an opportunistic large array (OLA) like structure to communicate the information broadcasted by any source to its intended receiver. The CWoT mechanism is found to be efficient in terms of QoS parameters for the adhoc networks in terms of reliability, energy efficiency and delay issues.

3 Cooperative web of trust (CWoT) for cognitive radio networks

The proposed CWoT security mechanism considers following model, which is a part of the Cooperative Opportunistic Large Array (OLA), as shown in Figure 4. The model illustrates various layers. The coverage limits of the various layers of the cooperation are shown with different levels. The analytical model for cooperative opportunistic large array (OLA) approach is considered same as in the works of same authors in [14]. Accordingly, the consumer radio devices (secondary user's sensor nodes) which are half-duplex in nature are assumed to be uniformly and randomly distributed over a continuous area with average density ρ. As in [10], the deterministic model is assumed, which means that the power received at a Consumer Premise Equipment (CPE) is the sums of powers from each of the CPE. In this model, the network node transmissions are orthogonal. It is assumed that a CPE can decode and forward a message without error when it's Signal to Noise ratio (SNR) is greater than or equal to modulation-dependent threshold λ_d. Due to noise variance assumption of unity, SNR criterion is transformed into received power criteria and λ_d becomes a power threshold. Let P_s be the source transmit power and the relay transmit power be denoted by P_r, and the relay transmit power per unit area be denoted by $\overline{P_r} = \rho P_r$. Instead of infinite radius, we are considering some practical scenarios where the radius is limited.

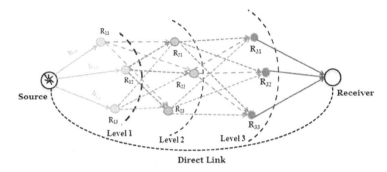

Figure 4 Proposed CWoT Model for Secondary users of CRN

Theoram: If $\mu \triangleq e^{(\lambda/\pi\rho P_R)}$ [15] and $\mu > 2$, then

$$r_k = \sqrt{\frac{P_s(\mu - 1)}{\lambda(\mu - 2)}} \left(1 - \frac{1}{(\mu - 1)^k}\right) \tag{1}$$

$$and \qquad \lim_{k\to\infty} r_k = r_\infty = \sqrt{\frac{P_s(\mu - 1)}{\lambda(\mu - 2)}} \tag{2}$$

For ($\mu \leq 2$), the broadcast reaches to the whole network i.e. $\lim_{k\to\infty} r_k = \infty$.

$$FES = 1 - \frac{number\ of\ active\ radio\ nodes utilized\ for\ cooperative\ transmission}{Total\ number\ of\ nodes\ in\ the\ OLA\ network}$$

For ($\mu > 2$), the total area reached by the broadcast is limited i.e. $r_k < r_{total}$ where r_k = radius of the kth level of the OLA structure.

Some preliminary assumptions for the proposed system are as below:

- All the nodes will have a unique identification, or UID, which will be utilized in the authentication of the nodes.
- All nodes are capable of transmitting and receiving information or data, if the minimum threshold for the received message is satisfied.

Taking into account the typical flow of the messages using the RTS-CTS-Message-ACK, the information about the nodes with the authentication details is transmitted cooperatively to the destination. The messages being relayed by the intermediary nodes or relays are considered on the basis of decode and forward, since the other technique amplify and forward amplifies the noise, thus degrading the signal that is received at the other end. It is in general considered that such cooperative relay of messages may present a problem of message flooding in the network. This situation is normally avoided by restricting the transmission of messages that fall below a given criteria (received SNR threshold) for the signal-to-noise ratio (SNR) of that message, as explained by [14]. The noise variance is assumed to be unity and hence the SNR criterion in transformed into a minimum criteria for power. Hence if the power with which the message is received is less than the threshold λ_t, the corresponding secondary relay node is not eligible for further retransmission of the signal. Such node stays idle during the communication.

3.1 Authentication

Let us consider an array of n nodes, depicted by N_i for $i = 1$ to n. Whenever a node, say N_A wants to communicate with N_B, N_A will send a Request to send (RTS) to N_B. Depending on whether the nodes are communicating for the first time or not, two scenarios are generated as explained below.

Scenario 1: *This is the first time that N_A is communicating with N_B*: When N_A is communicating for the first time with N_B, it would require an external entity to assure the authenticity of the node. The proposed model assumes that the network nodes trust each other to some basic level at the beginning of the communication, and later verifies the credibility of each node using trust values from other nodes. The basis of web-of-trust is used. For the aforementioned scenario i.e. if the nodes are communicating for the first time, some trust is to be assumed. In such a situation there is no way for N_A to verify that the person claiming to be N_B really is N_B. Hence N_A will, for the time being, trust N_B for the communication. An Asymmetric key exchange mechanism is considered. The public key is known to all the nodes in the network, whereas individual private key is retained by the corresponding node. Newer key exchange mechanisms for 802.11ae and 802.11af, based on groups have been discussed in the research work published in [15, 16, 17].

When N_A wants to communicate with N_B, it will use the public key of N_B, K_{public}, and will pass this, with its own UID to a one way function, F ($K_{public,}$UID). One way function is generated as shown in Figure 5. The output of this function, **G**, will then be sent to N_B. The use of one way function is beneficial as follows: any node other than N_B, will not be able to decipher the UID of N_A because of the use of the one way function. This is then attached to the RTS frame, which is to be sent to N_B. N_B, upon receipt of this frame recognizes that this is the first time N_A, or, for that communication, someone claiming to be N_A, is communicating with him/her.

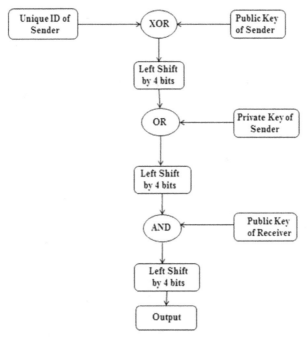

Figure 5   One Way Function Generation

Since there is no previous record of the authenticity of the identity of N_A, N_B will flag this node, and will try to confirm its identity later, as and when possible with cooperation from neighbouring nodes. This value G received from N_A is then given to another function F_D which will generate a corresponding value for the UID, called as K. Note that the UID itself is never disclosed to any other node. This value, K, is then stored in the memory of N_B. Based on feedback about this node from the neighbouring nodes, N_B may at a later stage delete this node, or set it to a higher priority.

In the packets that will follow, i.e. the ones containing the actual data from N_A, all N_B has to do is to extract the value of K of the sender from these packets. It may be noted that this value of K will be in encrypted format, if possible using the one way function only. N_B then extracts the K of the sender and matches it with K that it has received from the RTS packet. If the two keys match, the packet is considered as authentic and an acknowledgement is sent back to N_A. In the case that the value of K of the sender and the received packet do not match, the packet is discarded, with no notification being sent to the sender.

Scenario 2: N_A or someone claiming to be N_A has already communicated with N_B: In such a scenario, N_B has an idea about the identity of N_A. So all that N_B has to do is to confirm a match between the stored value of K of N_A and the value of K derived from the incoming packet. If a match occurs, the packets are processed and an acknowledgement is sent, otherwise the packet is discarded.

3.2 Trust Building

Cooperation itself has offered many of its benefits in the field of communication. The overheads of authentication can be reduced with the help of the cooperation from neighbouring nodes, i.e. by maintaining a web of trust (WoT). Consider a situation where N_A is a known party to N_B, i.e. they both trust each other. In a situation, where a third node, say Cairn, wants to contact with N_B. It is also known that N_A knows Cairn, that is, N_A trusts Cairn. This fact can be used to avoid unnecessary expenses that would be required to authenticate Cairn. As N_B trusts N_A, and N_A trusts Cairn, then a direct relation that N_B trusts Cairn can be made. Here, N_A is standing as a guarantor for Cairn.

It may be noted that this cooperation comes with its own drawbacks. Consider a situation where one of the nodes in the system is malicious. If this node stands as a guarantor for many other malicious nodes, then the security of the system can be compromised. One solution to this problem can be the use of trust-ranking of nodes. Based on the performance of nodes, ranks can be assigned to the nodes. If a node is a suspect, that is if many packets being sent via that node are not being delivered to the destination and this fact can be confirmed by some cooperation, then the node can be blacklisted, or its rank can be decreased by one. If the rank of a node reaches zero, its entry of K and node name is deleted and cooperatively notified to other nodes. If such a node has a guarantor, then the guarantor can be blacklisted and its priority be decreased as well. In the above example, N_A stood as a guarantor for Cairn. In the event that packets being routed through Cairn are not reaching the destination, or for that matter, if any crooked activity is suspected at Cairn then cairn can either be blacklisted or its rank can be decreased, depending upon the seriousness of the malicious nature being observed at that particular node. The message building mechanism uses the one way function as shown in Figure 6. The complete CWoT security mechanism is depicted as in the flowchart shown in Figure 7.

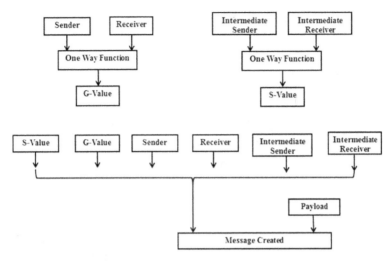

Figure 6   Message Creation Mechanism

In the event that a particular node has come up and is interacting with other nodes for the first time, the authenticity can be established based on the fact that if the new node is giving good performance with most number of neighbouring nodes, with no problems with the identity of that node, the node's rank can be increased, indicating the increased level of trust. We can do one more thing that trusted nodes can be added only at level 1, that is,

- A trusts B & B trusts C then A trusts C
- A trust B, B trusts C & C trust D then A trust D is not possible in this case.

In this scenario, we are assuming that by reducing the number of middle agents (secondary relays) will help us in improvement of security protocol.

3.3 Authorization

The role based access control technique is considered, wherein the participant radio nodes are classified according to various roles assigned to them. After the message is received, based on the reputation of that node, appropriate trust level is assigned to it. Based on the achieved trust level, the role is assigned to that particular node. Lastly, the access rights for that node are validated. The system divides the communicating nodes into three types of roles: sender, relay and receiver. Depending on the amount of information required for successful transmission of the message, appropriate access rights are assigned to these

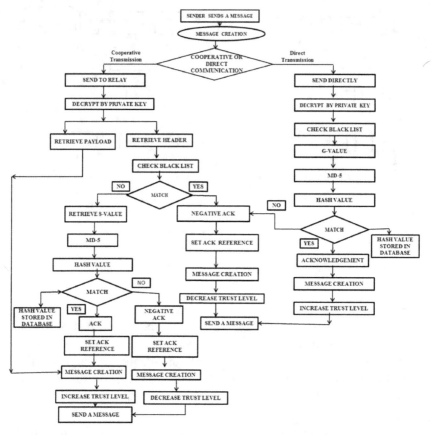

Figure 7   Flowchart for Cooperative Web of Trust (CWoT) Security Mechanism

roles. Though this scheme may look suitable for implementation, an obvious drawback of the previously implemented mechanisms is the static nature of the access roles that is provided to the participants. Therefore the roles are desired to be flexible. This can be achieved by the use of reputation based role assignment, as defined in [18]. In contrast to the multi-level approach of the technique proposed in the previous works, a decentralized approach is utilized here because of highly mobile nature of the nodes in the WSN. This gives equal priority to all the nodes, and reduces the central point of failure. On the basis of the trust level of the node that is communicating, a role is assigned to the node. It must be noted that since the role is being assigned at the necessary host, one node may have many roles assigned to it in context with different nodes. This may be thought of as a problem, but such a problem is easily eliminated as the trust information of the nodes is shared by all communicating parties. Thus, the trust value maintains appropriate reputation of the nodes, which in turn provides the suitable role to the node.

Proposed CWoT security system works for the detection and isolation of malicious nodes, based on the distance estimation of the values generated by broadcasting nodes, and gathering information about the same signal from neighbouring nodes. It is assumed that in replayed messages if the data that is presented, i.e. distance is incorrect and if such fact is brought to the notice of the node by the neighbours, then the nodes may diagnose it as a malicious node and thus eliminate it as shown in the authorization process of Figure 8.

In case, the identity of the adversarial relay (eavesdropper) is not diagnosed, then it can be pinpointed for detection by mechanism proposed in [19]. However it may be noted that only adversarial relay can be detected using this mechanism. The method involves the inclusion of some symbols. Based on the key shared between the sender and the receiver, the key that is unknown to the relay nodes, some symbols are generated. These symbols are called as trace symbols. The function explained above for the generation of the values of K (G value) can be used, along with some pseudo random number generator to produce unique values and the location where these symbols are to be added. At the receiving end, the receiver using the shared key extracts the symbol from the location. A mathematical function corresponding to the function used for the generation of the symbols is used at the receiving end to establish the ground truth whether these symbols were indeed generated on the basis of the tracing key, and then compare it with the received values. In case, the signal is garbled, or modified by a malicious relay, such malicious behaviour can be detected. Tracing mechanisms are provided in [20] for detection of the adversarial node.

The aforementioned topics give an insight to the basic mechanism that is to be implemented for this work. As every communication is bound to change the status of the network, it can be expressed as

$$M(n') \rightarrow < \alpha > M' \qquad (3)$$

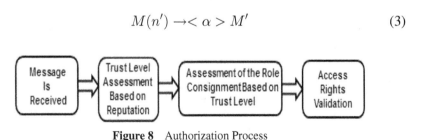

Figure 8 Authorization Process

When radio node of the network configuration transforms into another network configuration M' by execution of the action/communication/message. M is a table maintaining the trust information of the various participating nodes in the network. Each node stores the trust information about other nodes in its vicinity. A trust handling unit keeps on updating the trust level $T(n)$ of the neighbouring nodes. The trust levels can be categorized into ***blacklisted < not trusted < acquaintance < trusted < medium < highly trusted***. During initialization, the nodes are assigned a trust value of acquaintance. Thereafter, those are the communication messages that alter the trust level $T(n)$ of the neighbours. If a communication from node A to B delivers a corrupt message or the identity of the sender cannot be verified, it takes the network to a state that invokes decrease in the trust level of that node. As a node can act as the guarantor for other nodes that are less trusted, the guarantor stands liable for any false trust that it may have stood for.

This can be expressed as:

$$M(i) \rightarrow (\alpha)M' \tag{4}$$

$$M'(T(i) - 1) \tag{5}$$

Where M' = Broadcast trust

If the step (4) results into a trust of blacklist, that is $T(i) < \theta$, the node is removed or banned from communication. θ is the minimum value below which the node is blacklisted. There may be two scenarios existing after this case: (1) If it is observed that the blacklisted node is blacklisted by a node that is still trusted, that node's trust is decreased by one. (2) If the blacklisted node is blacklisted by many other networks, its trust level is decreased as well. Based on the trust $T(i)$ for the node i, roles are assigned. The role is represented as,

$$R(I, T(i)) \tag{6}$$

Where the role R is assigned to node i at trust level $T(i)$. The participants during communication will be assessed against this role at the receiving node. If it is found that the access requested is given in the role at the trust level $T(i)$, the action is permitted, otherwise rejected. The security is ensured as below: at the time of establishing communication, a trusted node at some trust level j will never communicate with another node at a trust level below some trust level k. This trust level k may vary from one node to another depending on the importance of the functionality of that node. In such manner, as proved by [21], malicious nodes are isolated from the communication network.

4 Simulation Results

After inclusion of security mechanisms in the communication system, it is general observation that the energy consumption of the system increases by large amount. As compared to the research work implemented in [15], the fraction of energy savings is slightly reduced with the addition of security in the system. As can be seen from the Figure 9, the energy consumption goes on increasing with the coverage area extension. It is interesting to note that for higher values of the SNR threshold (received SNR value at the secondary node), the energy consumption is observed to be reduced. The cooperative wireless communication is inherently energy efficient. By exploiting the cooperative diversity, the coverage range of the communicating nodes can be extended. Due to range extension capability, the received signal strength values are observed to be considerably better values compared to that without cooperation. This is very encouraging result for the protection against primary user emulation attack.

Received signal strength (RSS) at the secondary relay nodes is depicted in the Figure 10 below. It can be clearly seen that as compared to without cooperation, the RSS value is much better with cooperation. At the coverage radius of 30 meters, the RSS value is almost zero without cooperation whereas at the same value, the RSS value is found to be around 0.14 with cooperation. Secondary users can recognize each other's RSS signals and share a common protocol and are able to identify each other. Also due to increase or decrease in the trust levels due to the behaviour in the cooperative system, the secondary users are unable to emulate primary users. If any of the secondary user tries to misbehave and emulate primary user, its trust level gradually decreases and at the last, the node is blacklisted from the total communicating network entities.

The RSS value with cooperation is promising figure for the secondary entities in the cognitive radio networks. As in [6], the fraction of energy saving (FES) is given by,

$$FES = 1 - \frac{number\ of\ active\ radio\ nodes\ utilized\ for\ cooperative\ transmission}{Total\ number\ of\ nodes\ in\ the\ OLA\ network} \tag{5}$$

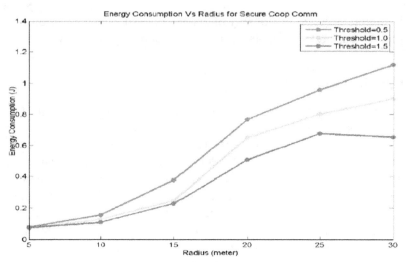

Figure 9   Energy Consumption Vs Coverage Radius for Secure Cooperative CRN

It is clearly observed from the figure that the FES value with security mechanism differs from the Cooperative system without security by almost 10%. Since the proposed system makes use of light weight cryptography and cooperative web of trust, the cost for the cooperative web of trust mechanism inclusion is less almost 10% as shown in Figure 11. Security inclusion cost of 10% is the promising result. Because the traditional cryptographic

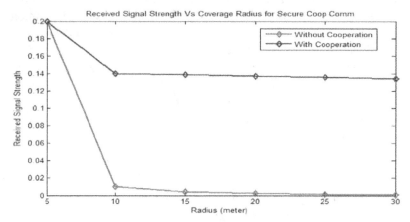

Figure 10   Received Signal Strength vs. Coverage Radius for Secure Web of Trust with and without Cooperation

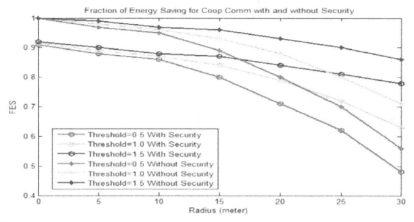

Figure 11   Fraction of Energy Savings Vs Coverage Radius with and without application of Secure Web of Trust

techniques are very much costly in terms of energy and computing power. Also, it is interesting to note that for higher values of threshold, the fraction of energy savings is considerably higher as compared to the low threshold situations.

5 Conclusions and Future Scope

The cooperative web of trust seems to provide promising energy efficient security solution for the spectrum sensing technique in cognitive radio networks. Also, the RSS values obtained are observed to be the effective result in the direction of the energy detection mechanism for spectrum sensing. Due to web of trust mechanism with cooperative diversity provides appropriate security solution for the primary user emulation attacks. Depending on the trust levels acquired through reputation in the system, the nodes immediately get either rewards for good behaviour or get blacklisted due to extreme misbehaviour. However, some improvements are needed in the present system. The storage of hash values is also a resource consuming prospect. Using proper function by light weight cryptography, the hash values can be computed at the run time, without consuming much time, thus eliminating the overheads of space and time requirements. Also since each broadcast consumes some energy, only relevant acknowledgements should be propagated, so that the system assumes the presence of an end-to-end logical channel, without having to bother about the intermediaries and the overhead such as acknowledgement sending to them. The authorization implemented assigns the role dynamically on the basis of reputation of the node.

References

[1] J. Mitola. Cognitive radio architecture evolution, Proceedings of IEEE Journals and Magazines, vol. 97, Issue 4, pp. 626–641, Apr. 2009.

[2] J. L. Burbank. Security in Cognitive Radio Networks: The Required Evolution in Approaches to Wireless Network, 3rd International conference on Cognitive Radio Oriented Wireless Networks and Communications, CrownCom 2008, pp. 1–7.

[3] K-C Chen et al. Cognitive radio network architecture: part I—general structure, In the Proceedings of the 2nd international conference on ubiquitous information management and communication, Suwon, Korea, 2008a, pp.114–119.

[4] S. Parvin, F. Khadeer Hussain, O. K. Hussain, S. Han, B. Tian, E. Chang. Cognitive radio network security: A survey, Journal of Network and Computer Applications, Vol.35, Issue. 06, November 2012, pp. 1691–1708.

[5] MIB Shahid, J. Kamruzzaman. Weighted soft decision for cooperative sensing in cognitive radio networks", 16th IEEE international conference on networks (ICON), New Delhi, 2008, pp. 1–6.

[6] A. M. Wyglinski, M. Nekovee, Y. T. Hou. Cognitive Radio Communications and Networks: Principles and Practice", Elsevier, Dec 2009.

[7] O. Leon, J. Hernandez-Serrano and M. Soriano. Securing Cognitive Radio Networks, International Journal of Communication Systems, Wiley InterScience, vol. 23, pp. 633–652, 2010.

[8] W. Wang, Y. (L.) Sun, H. Li and Z. Han. Cross-Layer Attack and Defense in Cognitive Radio Networks, IEEE Global Telecommunications Conference (GLOBECOM 2010), pp. 1–6, 2010.

[9] S. Parvin and F. K. Hussain. Trust-Based Security for Community Based Cognitive Radio Networks, 26th IEEE Conference on Advanced Information Networking and Applications, pp. 518–525, 2012.

[10] R. Dubey, S. Sharma, and L. Chouhan. Secure and Trusted Algorithm for Cognitive Radio Networks, Ninth International Conference on Wireless and Optical Communication Networks (WOCN), pp. 1–7, 2012.

[11] S. Parvin, S. Han, F. K. Hussain, Md. A. A. Faruque. Trust Based Security for Cognitive Radio Networks, Proceedings of the 12th International

Conference on Information Integration and Web-based Applications & Services, pp 743–748, 2010.

[12] H. Rifà-Pous, C. Garrigues. A Secure and Anonymous Cooperative Sensing Protocol For Cognitive Radio Networks, ACMSIN'11, Proceedings of the 4th international conference on Security of information and networks, pp 127–132.

[13] I. F. Akyildiz, Brandon F. Lo, Ravikumar Balakrishnan. Cooperative spectrum sensing in cognitive radio networks: A survey", Elsevier Science Direct Physical Communication 4, 2011, pp. 40–62.

[14] V. Rohokale, N. Kulkarni, N. Prasad, H. Cornean. Cooperative Opportunistic Large Array Approach for Cognitive Radio Networks, 8th International conference on communications, COMM 2010, pp. 513–516.

[15] W. Guo-feng, Z. Shi-lei, H. Xiao-ning and H. Han-Ying. A Trust Mechanism-based Secure Cooperative Spectrum Sensing Scheme in Cognitive Radio Networks", ESEP 2011: 9–10 December 2011, Singapore.

[16] J. Hong, Y. Qing-song, L. Hui. Simulation and Analysis of MAC Security Based on NS2, IEEE International Conference on Multimedia Information Networking and Security, vol.2, pp. 502–505, 2009.

[17] S. Misra, A. Vaish. Reputation-based role assignment for role based access control in wireless sensor networks, Computer Communications 34 (2011), pp.281–294, 2011.

[18] J. U. Duncombe. Infrared navigation—Part I: An assessment of feasibility (Periodical style), IEEE Tranactions on Electron Devices, vol.11, pp. 34–39, Jan. 1959.

[19] D. Liu, P. Ning. Security for Wireless Sensor Networks, Book Series in Advances in Information Security, Springer, ISBN 978-0-387-46781-8, vol.28, 2007.

[20] Y. Mao, M. Wu. Tracing Malicious Relays in Cooperative Wireless Communications, IEEE Transactions on Information Forensics and Security, Vol. 2, No. 2, pp. 198–212, June 2009.

[21] M. Merro and E. Sibilio. A Caluculus of Trustworthy Ad-hoc Networks", Springer-Verlag Berlin Heidelberg, pp. 157–172, 2010.

Biographies

Vandana Milind Rohokale received her B.E. degree in Electronics Engineering in 1997 from Pune University, Maharashtra, India. She received her Masters degree in Electronics in 2007 from Shivaji University, Kolhapur, Maharashtra, India. She is presently working as Assistant Professor in Sinhgad Institute of Technology, Lonavala, Maharashtra, India. She is currently pursuing her Ph.D. degree in CTIF, Aalborg University, Denmark. Her research interests include Cooperative Wireless Communications, AdHoc Networks and Cognitive Networks, Physical Layer Security, Information Theory and its Applications.

Dr. Neeli Prasad is leading a global team of 20+ researchers across multiple technical areas and projects in Japan, India, throughout Europe and USA. She has a Master of Science degree from Delft University, Netherlands and a PhD degree in electrical and electronic engineering from University of Rome Tor Vergata, Italy. She has been involved in projects totaling more than $120 million – many of which she has been the principal investigator. Her notable accomplishments include enhancing the technology of multinational players including Cisco, HUAWEI, NIKSUN, Nokia-Siemens and NICT as well as defining the reference framework for Future Internet Assembly and being one of the early key contributors to Internet of Things. She is also an advisor to the European Commission and expert member of governmental working groups and cross-continental forums. Previously, she has served as chief architect on large-scale projects from both the network operator and vendor side looking across the entire product and solution portfolio covering wireless, mobility, security, Internet of Things, Machine-to-Machine, eHealth, smart cities and cloud technologies. She has more than 250 publications and published two of the first books on WLAN. She is an IEEE senior member and an IEEE Communications Society Distinguished Lecturer.

Prof. Dr. Ramjee Prasad is the Director of the Center for TeleInfrastruktur (CTIF) and Professor Chair of Wireless Information Multimedia Communication at Aalborg University (AAU), Denmark. He is a Fellow of the Institute of Electrical and Electronic Engineers (IEEE), USA, the Institution of Electronics and Telecommunications Engineers (IETE), India; the Institution of Engineering and Technology (IET), UK; and a member of the Netherlands Electronics and Radio Society (NERG), and the Danish Engineering Society (IDA). He is recipient of several international academic, industrial and governmental awards of which the most recent is the Ridder in the Order of Dannebrog (2010), a distinguishment awarded by the Queen of Denmark.

Ramjee Prasad is the Founding Chairman of the Global ICT Standardisation Forum for India (GISFI: www.gisfi.org) established in 2009. GISFI has the purpose of increasing the collaboration between Indian, Japanese, European, North-American, Chinese, Korean and other worldwide standardization activities in the area of Information and Communication Technology (ICT) and related application areas. He is also the Founding Chairman of the HERMES Partnership (www.hermes-europe.net) a network of leading independent European research centres established in 1997, of which he is now the Honorary Chair.

Ramjee Prasad is the founding editor-in-chief of the Springer International Journal on Wireless Personal Communications. He is member of the editorial board of several other renowned international journals and is the series editor of the Artech House Universal Personal Communications Series. Ramjee Prasad is a member of the Steering, Advisory, and Technical Program committees of many renowned annual international conferences, e.g.,

Wireless Personal Multimedia Communications Symposium (WPMC); Wireless VITAE, etc. He has published more than 25 books, 750 plus journals and conferences publications, more than 15 patents, a sizeable amount of graduated Ph.D. students (over 60) and an even larger number of graduated M.Sc. students (over 200). Several of his students are today worldwide telecommunication leaders themselves.

Performance Evaluation on 6LoWPAN and PANA in IEEE 802.15.4g Mesh Networks

Yoshihiro Ohba[1] and Stephen Chasko[2]

[1]*Toshiba Corporate R&D Center, Japan*
[2]*Landis+Gyr, USA*
Corresponding Authors:
yoshihiro.ohba@toshiba.co.jp
stephen.chasko@landisgyr.com

Received 19 February 2014; Accepted 05 March 2014;
Publication 18 March 2014

Abstract

In this paper, we evaluate performance of 6LoWPAN and PANA over a mesh network based on mathematical analysis, considering IEEE 802.15.4 MAC/PHY behavior, 6LoWPAN behavior including fragmentation and PANA protocol behavior. End-to-end IP packet error rate, mean end-to-end IP packet delay, PANA session failure rate and mean PANA session establishment delay are used as the performance criteria. We show tradeoff points between Long Frame and Short Frame profiles for 6LoWPAN and PANA performance. As a result of performance analysis, we show a recommended PANA profile for IEEE 802.15.4g mesh networks to use Long Frame profile as long as MAC performance metric meet certain criterion.

Keywords: IEEE 802.15.4, 6LoWPAN, PANA, Fragmentation, Performance evaluation.

1 Introduction

Wireless mesh networks continue to deploy throughout the Smart Grid especially for HAN (Home Area Network) and NAN (Neighborhood Area Network) where efficient network devices and efficient operations of the devices are needed. ZigBee IP [1] is a IPv6-based network stack profile over

Journal of Cyber Security, Vol. 2 No. 3 & 4, 329–350.
doi: 10.13052/jcsm2245-1439.237

IEEE 802.15.4 [2] wireless mesh network for HAN. There is similar work underway to establish a profile for not only the HAN but for the NAN.

ZigBee IP uses PANA (Protocol for carrying Authentication for Network Access) [3] for network access authentication by transporting EAP (Extensible Authentication Protocol) [4], in conjunction with PANA relay extension [5] that is required for PANA to operate over multi-hop networks.

PANA is designed to be independent of link-layer technologies and is proven to be interoperable over small-scale ZigBee IP HANs that typically have only one or two hops between an end-device and the ZigBee IP coordinator. There has been no study on PANA for a large-scale IEEE 802.15.4 NAN. In order to define a NAN profile, it is important to clarify operational conditions of PANA for meta-networks that are common within the NAN.

The main goal of this document is to evaluate performance of 6LoWPAN (IPv6 over Low-Power Wireless Personal Area Networks) [6] and PANA over large-scale IEEE 802.15.4 mesh networks and define a PANA profile that works for NAN environments. In this paper, we develop a mathematical model for 6LoWPAN and PANA over a mesh network to evaluate the following performance metrics: end-to-end IP packet error rate and mean end-to-end IP packet delay, PANA session failure rate and mean PANA session establishment delay. Accuracy of the mathematical model for 6LoWPAN performance analysis is validated by simulation. We show tradeoff points between Long Frame and Short Frame profiles for 6LoWPAN and PANA performance.

Based on the performance evaluation, we show a recommended PANA profile GFSK (Gaussian Frequency-Shift Keying) PHY based IEEE 802.15.4g mesh networks to use Long Frame profile as long as MAC performance metric meet certain criterion. We further explore a cross-layer mechanism for dynamically changing fragment size taking into account not only the PHY and MAC profiles and performance metrics but also the application layer profiles and performance metric.

2 Network Model

The following network model is used. See also Figure 1. All nodes in the same IEEE 802.15.4g mesh network support IPv6 and 6LoWPAN for encapsulating IPv6 packets over 802.15.4 MAC. Each IEEE 802.15.4g MAC PDU (i.e., a MAC frame) can carry up to 2000 octets of MAC SDU. GFSK PHY with the maximum link speed of 100kbps is used where 1 symbol is equal to 1 bit. It is assumed that the channel is idle when an ACK frame is sent. A route-over mesh routing protocol such as RPL (IPv6 Routing Protocol for Low-Power

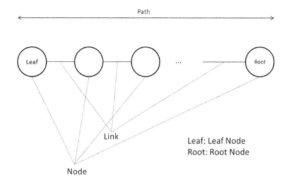

Figure 1   Network model

and Lossy Networks) [7] is used in the mesh network. The mesh network coordinator is referred to as a root node and the node on the other end of the PAN is referred to as a leaf node. The leaf node or the root node is the source node of 6LoWPAN packets. When the leaf node is the source node, the root node is the destination node, and vice versa. The path from the leaf node to the root node (i.e., the forward path) and the path from the root node to the leaf node (i.e., the reverse path) are symmetric. Hereafter the forward path and the reverse path are referred to as the path without distinction. There are H links along the path, constituting an H-hop path. 6LoWPAN header compression is not used. Fragmentation threshold for fragmenting an IP packet into multiple 6LoWPAN packets may be changed per packet and per hop, but does not change among multiple 6LoWPAN packets belonging to the same IP packet. The leaf node is the PaC and the root node is the PAA. For simplicity, we describe a model without utilizing the PANA relay element. On the other hand, our analysis can consider the impact of PANA relay by increasing the message size of each PANA message.

The following messaging model is used. See also Figure 2. The PaC initiates the PANA session. An authentication and authorization phase of a PANA session consists of an initiation followed by T transactions where T depends on the EAP authentication method in use. A successful initiation triggers the 1-st transaction. A successful i-th transaction triggers the $(i+1)$-st transaction. A PANA session is established if initiation and all T transactions are successful. An initiation consists of a PCI (PANA-Client-Initiation) message which is sent by the PaC. The PCI message will be retransmitted if the 1-st transaction does not start in a certain amount of time. The initiation is considered successful if the PCI message is received by the PAA before

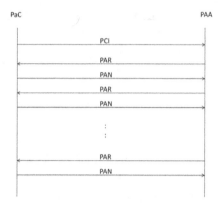

Figure 2   PANA messaging model

the number of retransmissions reaches its maximum value, R. A transaction consists of a PAR (PANA-Auth-Request) message sent by the PAA and a PAN (PANA-Auth-Answer) message sent by the PaC in response to the PAR. The PAR message will be retransmitted if the PAN message is not received in a certain amount of time. The transaction is considered successful if the PAN message is received by the PAA before the number of retransmissions reaches R. It is assumed that retransmissions of MAC frames of a PANA message in the network complete before the PANA retransmission timer for the PANA message expires.

3 Analysis

The notations used in the analytical model are shown in Table 1.

3.1 6LoWPAN Analysis

We analyze performance of IP over 6LoWPAN over an IEEE 802.15.4g mesh network employing un-slotted CSMA/CA with use of ACK frame (Figure 3). The sender of a data frame waits for LIFS (Long Inter-Frame Space) seconds after the last received ACK frame and before starting CSMA/CA back-off for data frame transmission. The sender of an ACK frame waits for SIFS (Short Inter-Frame Space) seconds after receipt of a data frame and before transmitting the ACK frame. The CSMA/CA back-off algorithm used in IEEE 802.15.4 MAC is shown in Figure 4.

   1) End-to-end Packet Error Rate

First, $f_{l,tx}(L)$ is computed from f_b and $e_d(L)$ as $f_{l,tx}(L) = f_b + (1 - f_b) e_d(L)$, where $f_b = 1 - \sum_{j=0}^{B_{\max}} c^j (1 - c) = c^{B_{\max}+1}$.

Table 1   Notations

Name	Meaning
	MAC and PHY Parameters
L	Data frame size in octets.
L_a	ACK size in octets. $L_a = 4$ octets in IEEE 802.15.4.
M	Maximum number of retransmissions of link-layer frame.
$B_{\max}$	The maximum number of CSMA/CA back-offs (default = 4).
u	Back-off unit. $u = 20$ (bits) for GFSK PHY.
$E_{\min}$	Minimum back-off exponent (default = 3).
$E_{\max}$	Maximum back-off exponent (default = 5).
d_{AW}	ACK wait time. $d_{AW} = 6u/c$ for GFSK PHY..
d_{BO}	Mean CSMA/CA back-off time. $d_{BO} = (2^{E_{\min}} - 1)u/2/c = 70/c.$
d_{LIFS}	Inter-frame spacing latency for data. $d_{LIFS} = 40/c$ for GFSK PHY.
d_{SIFS}	Inter-frame spacing latency for ACK. $d_{SIFS} = 12/c$ for GFSK PHY.
d_{PROC}	Frame processing latency for data. We assume $d_{PROC} = 0.0$.
C	Link speed in bps. $C = 100000$ for GFSK PHY.
e	Bit error rate per link.
c	Channel busy rate.
	PANA Parameters
R	Maximum number of retransmissions of a PCI or PAR message.
T	Number of transactions. We use $T = 4$ which is a typical number for EAP-TLS.
m_0^-	The number of link-layer frames encapsulating a PCI message.
L_0	The frame length of each frame encapsulating a PCI message.
$m_{i,\text{req}}$	The number of link-layer frames encapsulating a PAR message in i-th transaction.
$L_{i,\text{req}}$	The frame length of each link-layer frame encapsulating a PAR message in i-th transaction.
$m_{i,\text{ans}}$	The number of link-layer frames encapsulating a PAN message in i-th transaction.
$L_{i,\text{ans}}$	The length of each link-layer frame encapsulating a PAN message in i-th transaction.
IRT_0	Initial retransmission interval in seconds for PCI message. We use $IRT_0 = 15$.
$IRT_{0,\max}$	Maximum retransmission interval in seconds for PCI message. We use $IRT_{0,\max} = 120$.
IRT_r	Initial retransmission interval in seconds for PAR message. We use $IRT_r = 10$.
$IRT_{r,\max}$	Maximum retransmission interval in seconds for PAR message. We use $IRT_{r,\max} = 30$.

(Continued)

Table 1 Continued

MAC Performance Metrics	
$e_d(L)$	Frame error rate for data frame of length L octets. $e_d(L) = 8Le$.
e_a	Frame error rate for ACK frame. $e_a = 8L_a e$.
f_b	CSMA/CA failure rate for data frame.
$f_{l,tx}(L)$	Failure rate for transmission of a MAC data frame of length L octets.
$f_{l,tr}(L)$	Failure rate for an exchange of a MAC data frame of length L octets and an ACK frame sent in response to the data frame.
6LoWPAN Performance Metrics	
$f_p(m, L)$	Packet transmission failure rate over the path for an IP packet consisting of m 6LoWPAN frames of length L octets.
$d_l(m, L)$	Mean per-hop transmission latency in seconds for an IP packet consisting of m 6LoWPAN frames of length L octets.
$d_e(m, L)$	Mean end-to-end delay in seconds for an IP packet consisting of m 6LoWPAN fragments of length L octets.
PANA Performance Metrics	
$e_r^{(i)}$	A failure rate of a PAR transmission in i-th transaction ($i > 0$).
$e_t^{(i)}$	A failure rate of i-th transaction.
e_S	A failure rate of PANA session establishment (i.e., PANA session error rate).
D	Mean PANA session establishment delay in seconds

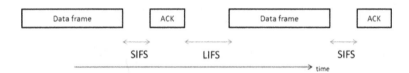

Figure 3 CSMA/CA with ACK

Based on the assumption that the channel is idle when ACK is sent, $f_{l,tr}(L)$ is calculated as $f_{l,tr}(L) = 1 - (1 - f_{l,tx}(L))(1 - e_a)$.

Since there are H links between the originating node and the destination node for an m–fragment IP packet, $f_p(m, L)$ is computed as $f_p(m, L) = 1 - \{(1 - f_{l,tr}(L)^{M+1})^{m-1}(1 - f_{l,tx}(L)^{M+1})\}^H$.

2) Mean End-to-end Packet Delay

Since the first $(m - 1)$ requires an ACK frame and the link-layer retransmission interval is d_{AW}, the back-off time is d_{BO}, and the MAC frame transmission latency is $8L/C$, and the transmission of an ACK frame for the last (i.e., m-th) data frame of an IP packet does not contribute to the latency

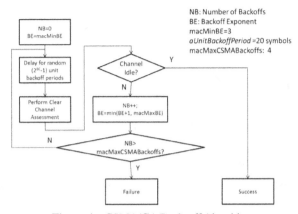

Figure 4   CSMA/CA Back-off Algorithm

of the IP packet for each hop, $d_l\,(m, L)$ is computed as follows.

$$d_l\,(m, L) = (m - 1) \times$$

$$\frac{\sum_{j=0}^{M}\left(\begin{array}{c}j\,(8L/C+d_{BO}+d_{AW})+8\,(L+L_a)/C+\\d_{BO}+d_{LIFS}+d_{SIFS}+d_{PROC}\end{array}\right)f_{l,tr}\,(L)^j\,(1-f_{l,tr}(L))}{(1-f_{l,tr}(L)^{M+1})^{m-1}}+$$

$$\frac{\sum_{j=0}^{M}(j(8L/C+d_{BO}+d_{AW})+8L/C+d_{BO}+d_{LIFS}+d_{PROC})f_{l,tx}(L)^j\,(1-f_{l,tx}(L))}{(1-f_{l,tx}(L)^{M+1})},$$

where

$$d_{\rm BO} = \sum_{j=0}^{B_{\max}}\left(2^{\min(j+E_{\min},\,E_{\max})} - 1\right)(u/2C)\,c^j\,(1 - c)$$

$d_e\,(m, L)$ is given by $d_e\,(m, L) = Hd_l\,(m, L)$.

3.2 PANA Analysis

1) Session Failure Rate

Since a single transmission of a PAR message can result in a successful receipt of a PAN message when both the request and answer messages are successfully transmitted, $e_r^{(i)}$ is computed as:

$$e_r^{(i)} = 1 - \{1 - f_p\,(m_{i,\text{req}}, L_{i,\text{req}})\}\{1 - f_p\,(m_{i,\text{ans}}, L_{i,\text{ans}})\}.$$

Since a transaction fails when all retransmissions of the PAR message fail, $e_t^{(i)} = \{e_r^{(i)}\}^{R+1}$.

Finally, a PANA session establishment fails when initiation fails or one of T transactions fails, e_s is computed as:

$$e_s = f_p(m_0, L_0)^{R+1} + \{1 - f_p(m_0, L_0)^{R+1}\}\{1 - \prod_{i=1}^{t}(1 - e_t^{(i)})\}.$$

2) Mean Session Establishment Delay

Let $d_k^{(i)}$ be the mean delay for i-th transaction that succeeds after k retransmissions. Then, for $i > 0$,

$$d_k^{(i)} = g_k \min\left(IRT_r 2^{k-1}, IRT_{r,max}\right)$$
$$+ H\left(d_e\left(m_{i,req}, L_{i,req}\right) + d_e\left(m_{i,ans}, L_{i,ans}\right)\right), \text{ where}$$

$$g_k = \begin{cases} 1, & k > 0 \\ 0, & k = 0 \end{cases}.$$

Let $d^{(i)}$ be the mean delay for i-th transaction, then $d^{(i)}$ is compute from $d_k^{(i)}$ as follows.

$$d^{(0)} = \frac{\sum_{k=0}^{R}\{g_k \min\left(IRT_0 2^{k-1}, IRT_{0,max}\right) + Hd_e(m_0, L_0)\}f_p(m_0, L_0)^k(1 - f_p(m_0, L_0))}{1 - f_p(m_0, L_0)^{R+1}}.$$

$$d^{(i)} = \frac{\sum_{k=0}^{R} d_k^{(i)}\{e_r^{(i)}\}^k(1 - e_r^{(i)})}{1 - e_t^{(i)}}, i > 0.$$

Finally, $D = \sum_i^{T} d^{(i)}$

Note that the computed D value is valid if the maximum roundtrip time is smaller than the initial retransmission interval. Therefore, the operational condition of the system is given by:

$$H < \frac{IRT_r C}{2m_{max}(8L)}, \text{ where } m_{max} = \max\left\{\max_i\{m_{i,req}\}, \max_i\{m_{i,ans}\}\right\}.$$

For example, $H \leq 28$ for $(m_{max}, L, IRT_r) = (17, 127, 10)$, and $H \leq 47$ for $(m_{max}, L, IRT_r) = (1, 1327, 10)$.

4 Performance Evaluation

We use two types of fragment size profiles, i.e., "Short Frame" and "Long Frame" as described in Table 2.

Table 2 Fragment Size Profile

6LoWPAN Parameter	Value	
	Short Frame	Long Frame
Maximum fragment size	127 octets	1327 octets

4.1 6LoWPAN Evaluation

First, we show 6LoWPAN performance in terms of packet loss rate given by $f_p(m, L)$ and mean delay given by $d_e(m, L)$ for an IPv6 packet of length 1280 octets, $M = 3$, and $H = 1, 2 \ldots 10$, where $(m, L) = (18, 127)$ for Short Frame profile and $(m, L) = (1,1327)$ for Long Frame profile . Results for 4 cases $(c, e) = (0.0, 0.00001)$, $(0.0, 0.00003)$, $(0.2, 0.00001)$, $(0.2, 0.00003)$ based on analysis and simulation are shown in Table 3, Table 4, Table 5, and Table 6, respectively. Tens of thousands of IP packets are generated in each simulation run. In the simulations, adjacent nodes are placed at a distance of 100m which is equal to the radio coverage. These tables indicate that difference in analysis and simulation results are close in the examined parameter range. Specifically, mean delay difference is within 5.2 % and packet loss difference is within 2 orders of magnitude for packet loss rate higher than 10^{-6} and within the 1 order of magnitude for packet loss rate higher than 10^{-4}. Hereafter our evaluation is based on analysis only.

Table 3 6LoWPAN Performance ($c = 0.0$, $e = 0.00001$, $M = 3$)

	Mean Delay			
	Short ($c = 0.0$, $e = 0.00001$, $M = 3$)		Long ($c = 0.0$, $e = 0.00001$, $M = 3$)	
H	Simulation	Analysis	Simulation	Analysis
1	0.206555	0.212455	0.119791	0.12054
2	0.413213	0.424909	0.240037	0.24108
4	0.826231	0.849818	0.479471	0.48216
6	1.239213	1.274727	0.715689	0.72324
8	1.652637	1.699636	0.95945	0.96432
10	2.065444	2.124545	1.197706	1.2054
	Packet Loss Rate			
	Short ($c = 0.0$, $e = 0.00001$, $M = 3$)		Long ($c = 0.0$, $e = 0.00001$, $M = 3$)	
H	Simulation	Analysis	Simulation	Analysis
1	0	2.15E-07	0.0012	0.00012894
2	0.0001	4.31E-07	0.0026	0.00025786
4	0	8.62E-07	0.0052	0.00051565
6	0.0001	1.29E-06	0.0075	0.00077337
8	0.0001	1.72E-06	0.0118	0.00103103
10	0.0001	2.15E-06	0.013	0.00128862

Table 4   6LoWPAN Performance (c = 0.0, e = 0.00003, M = 3)

	Mean Delay			
	Short (c = 0.0, e = 0.00003, M = 3)		Long (c = 0.0, e = 0.00003, M = 3)	
H	Simulation	Analysis	Simulation	Analysis
1	0.21072	0.21719	0.148083	0.154046
2	0.421656	0.43438	0.292793	0.308092
4	0.842965	0.868761	0.5877	0.616185
6	1.265098	1.303141	0.884022	0.924277
8	1.685981	1.737521	1.173841	1.232369
10	2.107599	2.171902	1.46821	1.540462
	Packet Loss Rate			
	Short (c = 0.0, e = 0.00003, M = 3)		Long (c = 0.0, e = 0.00003, M = 3)	
H	Simulation	Analysis	Simulation	Analysis
1	0.0004	1.7412E-05	0.0348	0.01044388
2	0.0007	3.4823E-05	0.0646	0.02077869
4	0.0015	6.9644E-05	0.1249	0.04112562
6	0.0031	0.00010446	0.1753	0.06104977
8	0.0028	0.00013928	0.2349	0.08055992
10	0.0035	0.0001741	0.2827	0.09966467

Table 5   6LoWPAN Performance (c=0.2, e = 0.00001, M = 3)

	Mean Delay			
	Short (c = 0.2, e = 0.00001, M = 3)		Long (c = 0.2, e = 0.00001, M = 3)	
H	Simulation	Analysis	Simulation	Analysis
1	0.214697	0.216583	0.12002	0.120828
2	0.42972	0.433165	0.240763	0.241656
4	0.859242	0.86633	0.480075	0.483311
6	1.288571	1.299496	0.72124	0.724967
8	1.718689	1.751807	0.961933	0.966623
10	2.147817	2.189758	1.203159	1.208278
	Packet Loss Rate			
	Short (c = 0.2, e = 0.00001, M = 3)		Long (c = 0.2, e = 0.00001, M = 3)	
H	Simulation	Analysis	Simulation	Analysis
1	0	2.4277E-07	0.0012	0.00013033
2	0	4.8553E-07	0.0019	0.00026064
4	0.0001	9.7106E-07	0.0045	0.0005212
6	0.0002	1.4566E-06	0.0073	0.0007817
8	0.0003	1.9421E-06	0.0119	0.00104213
10	0.0003	2.4277E-06	0.0154	0.0013025

Next we show packet loss rate and mean packet delay performance in broader set of parameters. Figure 5 and Figure 6 show packet loss rate versus H for $(c, e) = (0.1, 0.00001)$, $(0.1, 0.00003)$, respectively for $M = 3, 7$. Figure 7 and Figure 8 show mean packet delay versus H for

Table 6   6LoWPAN Performance ($c = 0.2$, $e = 0.00003$, $M = 3$)

	Mean Delay			
	Short (c = 0.2, e = 0.00003, M = 3)		Long (c = 0.2, e = 0.00003, M = 3)	
H	Simulation	Analysis	Simulation	Analysis
1	0.219133	0.221407	0.147527	0.154403
2	0.43853	0.442815	0.295205	0.308806
4	0.876995	0.88563	0.58945	0.617612
6	1.31556	1.328445	0.885841	0.926418
8	1.754518	1.77126	1.177391	1.235224
10	2.192648	2.214075	1.477293	1.544031
	Packet Loss Rate			
	Short (c = 0.2, e = 0.00003, M = 3)		Long (c = 0.2, e = 0.00003, M = 3)	
H	Simulation	Analysis	Simulation	Analysis
1	0.0006	1.811E-05	0.0351	0.01047236
2	0.0013	3.622E-05	0.0632	0.02083505
4	0.0022	7.2439E-05	0.1302	0.04123599
6	0.0029	0.00010866	0.1844	0.06121188
8	0.0058	0.00014487	0.2362	0.08077158
10	0.0061	0.00018109	0.2787	0.09992374

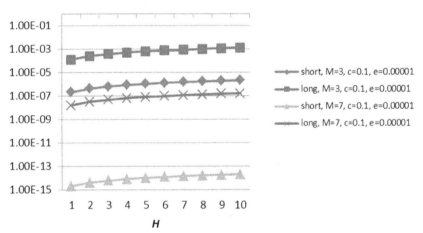

Figure 5   Packet Loss Rate vs. Number of Hops (c = 0.1, e = 0.00001)

$(c, e) = (0.1,\ 0.00001)$, $(0.1, 0.00003)$, respectively for $M = 7$. Note that mean packet delay values for $M = 3$ are nearly the same as those for $M = 7$. Short Frame shows more than 4 orders of magnitudes lower packet loss rate than Long Frame profile. On the other hand, Long Frame profile shows less than 1/2 the lower packet delay than Short Frame profile.

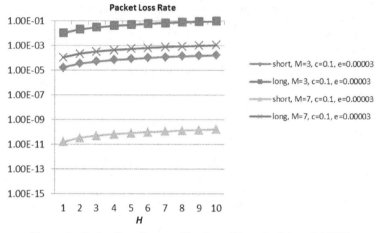

Figure 6   Packet Loss Rate vs. Number of Hops (c=0.1, e=0.00003)

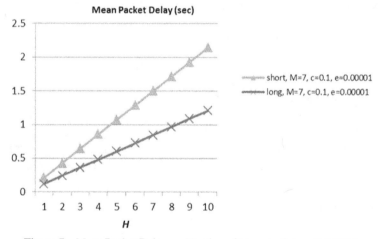

Figure 7   Mean Packet Delay vs. Number of Hops (c=0.1, e=0.00001)

Figure 9 and Figure 10 show packet loss rate and mean packet delay performance versus c, respectively, for $(H, e) = (10,\ 0.00001)$ and $M = 3, 7$. Again Long Frame profile shows less than ½ the lower packet delay than Short Frame profile. When channel busy rate is low, Short Frame profile shows smaller packet loss rate than Long Frame profile, but when channel busy rate exceeds a certain threshold, Short Frame profile incurs higher packet loss rate than Long Frame profile. This can be explained as follows. A MAC frame

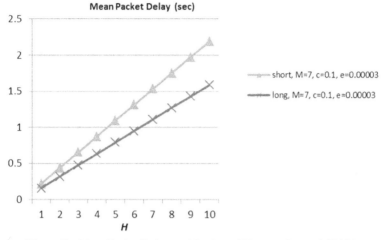

Figure 8   Mean Packet Delay vs. Number of Hops (c=0.1, e=0.00003)

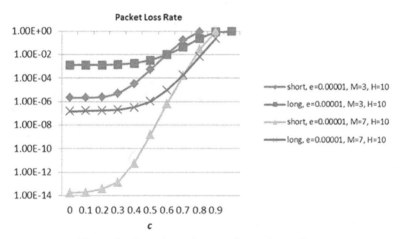

Figure 9   Packet Loss Rate vs. Channel Busy Rate

loss can be caused by bit errors and CSMA/CA back-off failures. The former occurs more frequently when the fragment size is large. On the other hand, the latter occurs more frequently when the channel busy rate is high. The benefit of smaller fragment size to be robust against bit errors is negated by the disadvantage of the larger number of smaller sized fragments which causes higher CSMA/CA back-off failure rate, and the disadvantage overwhelms the advantage where the channel busy rate exceeds the threshold.

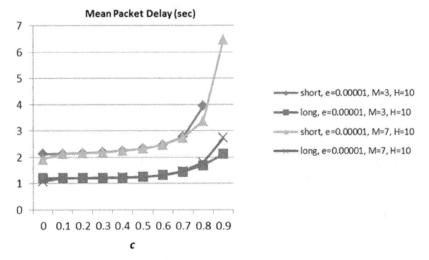

Figure 10   Packet Loss Rate vs. Channel Busy Rate

Table 7   PANA Profile

PANA Operational Parameters	Value	
	Short Frame	Long Frame
m_0	1 frame	1 frame
$m_{i,\text{req}}, m_{i,\text{ans}}\ (0 < i \le 4)$	16 frames	1 frame
L_0	127 octets	127 octets
$L_{i,\text{req}}, L_{i,\text{ans}}\ (0 < i \le 4)$	127 octets	1327 octets

4.2 PANA Evaluation

In this section we evaluate performance of PANA over 6LoWPAN in terms of packet loss rate given by e_S and mean session establishment delay given by D. We use the following profile for PANA. Both profiles are corresponding to PCI message of 80 octets in IP PDU length, followed by a sequence of $T{=}4$ pairs of PAR and PAN messages all of which have 1280 octets in IP PDU length.

Figure 11 and Figure 12 show PANA session failure rate versus M for $R = 1$ and $R = 5$, respectively for c = 0.0. Figure 13 and Figure 14 show PANA session establishment delay versus M for $R = 1$ and $R{=}\,5$, respectively for c = 0.0. Figure 15 and Figure 16 show PANA session failure rate and PANA session establishment delay versus c , respectively, for $H{=}\,10$, $M{=}\,7$, and $R{=}\,5$. The following observations can be made.

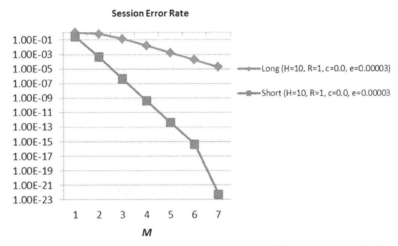

Figure 11   PANA Session Error Rate vs. Max. Number of MAC Frame Retransmission (*R*=1)

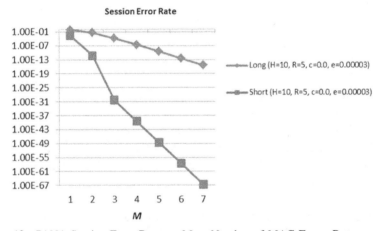

Figure 12   PANA Session Error Rate vs. Max. Number of MAC Frame Retransmissions (*R*=5)

Increasing the maximum number of MAC frame retransmissions can decrease both PANA session error rate and PANA session establishment delay for both Short Frame and Long Frame cases (Figure 11, Figure 12, Figure 13 and Figure 14). Increasing the maximum number of PANA message retransmissions can decrease PANA session error rate and increase PANA session establishment delay for both Short Frame and Long Frame cases (Figure 11, Figure 12, Figure 13 and Figure 14). Short Frame profile shows

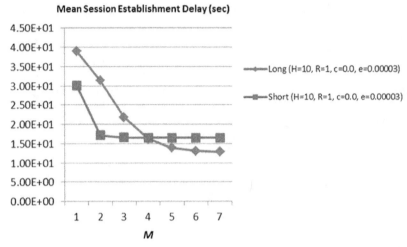

Figure 13   Mean PANA Session Establishment Delay vs. Max. Number of MAC Frame Retransmissions (*R*=1)

lower PANA session error rate than Long Frame profile when the channel busy rate is low (Figure 11 and Figure 12). Long Frame profile shows lower PANA session establishment delay than Short Frame profile when the maximum number of MAC frame retransmission is high (Figure 16). There is a threshold for the maximum number of MAC frame retransmissions below which PANA session establishment delay for Long Frame profile becomes larger than that for Short Frame profile when channel busy rate is low (Figure 13 and Figure 14). There is a threshold for the channel busy rate above which PANA session error rate for Short Frame profile becomes larger than that for Long Frame profile (Figure 15).

From these observations, it is recommended to use Long Frame profile defined in Section 0 for PANA with $M = 7$ and the following PANA session parameters for achieving PANA session error rate lower than 10^{-7} and mean PANA session establishment delay lower than 20 seconds in 6LoWPAN over IEEE 802.15.4g networks that employ GFSK PHY with the link speed (C) of 100 kbps with the number of hops (H) not exceeding 10 and bit error rate (e) not exceeding 0.00003 and channel busy rate (c) not exceeding 0.6: $R = 5$, $IRT_0 = 15$, $IRT_{0,\max} = 120$, $IRT_r = 10$ and $IRT_{r,\max} = 30$

In the case where bit error rate on the outgoing link of a node exceeds 0.00003 and channel busy rate does not exceed 0.6, it is recommended to either switch to use Short Frame profile or use an alternative next hop node.

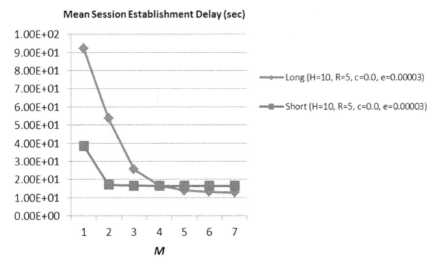

Figure 14 Mean PANA Session Establishment Delay vs. Max. Number of MAC Frame Retransmission (*R*=5)

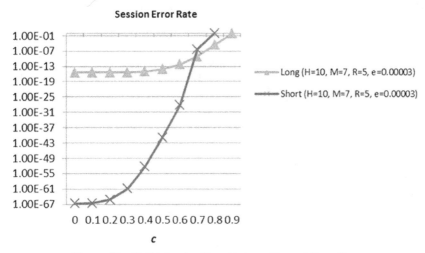

Figure 15 PANA Session Error Rate vs. Channel Busy Rate

In the case where channel busy rate on the outgoing link of a node exceeds 0.6, it is recommended to use an alternative next hop node without switching to Short Frame profile. A more simplified way is to use an alternative next hop node if packet loss rate bit error rate exceeds 0.00003 or channel busy rate exceeds 0.6.

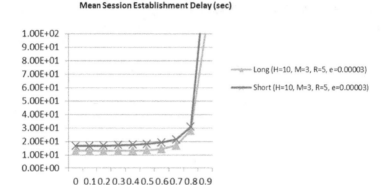

Figure 16   Mean PANA Session Establishment Delay vs. Channel Busy Rate

5 Guidelines on Fragment Size

Although the analysis and performance evaluation described in this paper is focused on PANA over IPv6 employing 6LoWPAN encapsulation, the obtained results lead to an insight into a cross-layer mechanism for dynamically changing fragment size taking not only the PHY and MAC profiles and performance metric but also application layer profiles and performance metric into accounts. As a rule of thumb, in the environments such as IEEE 802.15.4 wireless mesh networks where MAC frame loss rate is not ignorant, use of smaller 6LoWPAN fragment size is recommended for real-time applications since they do not rely on retransmissions above MAC layer, and use of larger 6LoWPAN size is recommended for non-real time applications that employs a transport or application layer retransmission mechanism that can compensate IP packet loss.

Note that this kind of dynamic fragment size control is applicable to IP-layer fragmentation as well. However, it is recommended to use 6LoWPAN fragmentation wherever available because 6LoWPAN provides not only fragmentation but also more efficient header compression schemes [8] than IP header compression [9].

6 Conclusions

In this document, we developed a mathematical model for 6LoWPAN and PANA over a mesh network to evaluate end-to-end IP packet error rate and mean end-to-end IP packet delay, PANA session failure rate and mean PANA session establishment delay.

Through performance evaluation of 6LoWPAN, we observed that Short Frame profile always shows larger mean packet delay than Long Frame profile, and when channel busy is low, Short Frame profile shows smaller packet loss rate than Long Frame profile. On the other hand, when channel busy rate exceeds a certain threshold, Short Frame profile incurs higher packet loss rate than Long Frame profile. Through performance evaluation of PANA over 6LoWPAN, we observed that increasing the maximum number of MAC frame retransmissions can decrease both PANA session error rate and PANA session establishment delay, and increasing the maximum number of PANA message retransmissions can decrease PANA session error rate at the cost of larger PANA session establishment delay. We also observed tradeoff points in terms of PANA session establishment delay as well as PANA session error rate between Long Frame and Short Frame profiles.

As a result, a recommended PANA profile was introduced for GFSK-based IEEE 802.15.4g mesh networks to use Long Frame profile as long as MAC performance metric meet certain criterion.

Finally, we explored an idea of cross-layer mechanism for dynamically changing fragment size taking not only the PHY and MAC profiles and performance metric but also application layer profiles and performance metric into accounts. We plan to investigate such a mechanism deeply in our future work.

7 Acknowledgment

The authors gratefully acknowledge the support of Mitsuru Kanda, Seijiro Yoneyama, Yasuyuki Tanaka, Mike Demeter and Ruben Salazar for this work.

References

[1] ZigBee Alliance, "ZigBee IP Specification", ZigBee Public Document 13–002r00, 2013.

[2] IEEE, "IEEE Standard for LAN/MAN—Specific requirements Part 15.4: Wireless Medium Access Control (MAC) and Physical Layer (PHY) Specifications for Low-Rate Wireless Personal Area Networks (WPANs)," 2011.

[3] Y. Ohba, a. et., "Protocol for Carrying Authentication for Network Access (PANA)," RFC 5191, 2008.

[4] B. Aboba and a. et., "Extensible Authentication Protocol (EAP)," RFC 3748, 2004.

[5] P. Duffy, S. Chakrabarti, R. Cragie, Y. Ohba and A. Yegin, Protocol for Carrying Authentication for Network Access (PANA) Relay Element, RFC 6345, 2011.

[6] G. Montenegro, N. Kushalnagar, J. Hui and D. Culler, Transmission of IPv6 Packets over IEEE 802.15.4 Networks, RFC 4944, 2007.

[7] E. T. Winter , E. P. Thubert, RPL: IPv6 Routing Protocol for Low power and Lossy Networks, RFC 6550.

[8] J. Hui and P. Thubert, "Compression Format for IPv6 Datagrams over IEEE 802.15.4-Based Networks," RFC 6282, 2011.

[9] M. Degermark, B. Nordgren and S. Pink, "IP Header Compression," RFC 2507, 1999.

Biographies

Yoshihiro Ohba is a Chief Research Scientist in Toshiba Corporate R & D Center, Japan. He received B.E., M.E. and Ph.D. degrees in Information and Computer Sciences from Osaka University in 1989, 1991 and 1994, respectively. He is an active member in IEEE 802 and IETF for standardizing security and mobility protocols. He is Chair of IEEE 802.21a Task Group and IEEE 802.21d Task Group, and Vice Chair and Secretary of ZigBee Alliance Neighborhood Area Network (NAN) WG. He is a main contributor to RFC 5191 (PANA - Protocol for carrying Authentication for Network Access). He received IEEE Region 1 Technology Innovation Award 2008.

Stephen Chasko (M'1888, F'17) was born in San Diego, California on August 5, 1969. He graduated from the Arizona State University with a BSEE.

His employment experience includes NCR, ACI Worldwide, Texas Instruments. His special fields of interest include data communications, smart card systems, secure microcontrollers and smart grid security He is currently a Security+Software Manager at Landis+Gyr.

Mr. Chasko has spoken at numerous security conferences including the RSA Conference, Smart Grid Security, Distributech and IEEE PES.

NGSON Service Composition Ontology

Reinhard Schrage

SchrageConsult Seelze, Germany reinhard.schrage.de@ieee.org

Received 5 February 2014; Accepted 28 February 2014;
Publication 8 April 2014

Abstract

The paradigm of Next Generation Service Overlay Networks (NGSON) strives to provide a unified and standardized framework of IP-based service overlay networks, creating an ecosystem of context-aware, dynamically adaptive, and self-organizing networking capabilities, including advanced routing and forwarding schemes. Service composition, i.e. the facility to combine certain atomic services into an aggregated service is considered a vital part of NGSON. As a future directed, next generation oriented paradigm NGSON must enable an intelligent, automated service composition platform. In order to also meet the objectives of being context-aware, dynamically adaptive, and self-organizing this platform needs to know and understand the semantics of its underlying functional entities. Furthermore, to be accepted by users, enterprises and service developers existing, proven, but likewise extendable standards need to be utilized as much as possible.

The W3C consortium has released OWL2 for building ontologies that serve to provide machine-understandable semantics. In order to be feasible for NGSON a service composition ontology also needs to include concepts from deontic logic, i.e. needs to be able to differentiate between omissible and permissible classes, or -in finer granularity- prohibited, obligatory and optional components when composing a service from atoms.

This paper aims to underline the need for an OWL2 ontology, make suggestions on its structure and required interfaces to other network entities as e.g. Software Defined Networks and Network Virtualization Functions.

Keywords: NGSON; Service Composition; OWL2 Ontology.

Journal of Cyber Security, Vol. 2 No. 3 & 4 , 351–358.
doi:10.13052/jcsm2245-1439.238

1 Introduction

Already, there is a plethora of networking applications available. Yet, very often, these applications form independent silos unaware of each other's functionalities and unable to communicate and benefit from already existing and implemented algorithms. Users, on the other hand, find that they need to invoke several applications in order to retrieve all relevant information required for a particular query. Output from one application may have to be manually re-entered into another application as input, probably also in a different format. Software developers are faced with the necessity of adapting their applications due to minor changes in the rendering of a webpage used to collect input data.

It is evident that users and suppliers alike will benefit from a capability of data integration across individual applications. Semantic Web Services are addressing this horizontal data integration.

At the same time there are new developments in the design and implementation of transport networks that carry user applications: Network Functions Virtualization (NFV) aims at reducing CAPEX, OPEX, space and power consumption by hosting, i.e. virtualizing, dedicated single purpose components into general purpose servers, while Software Defined Networking decouples the control and data planes, centralizing network intelligence and state, and abstracting the underlying network infrastructure from the applications to facilitate faster innovation.

Amongst other issues a platform that is to facilitate machine-automated, context-aware and dynamically adaptive composition of service applications and provide rule-based optimization of transport streams needs to:

1. Know what services/applications are available at request time
2. Know how these services, if at all, can be composed

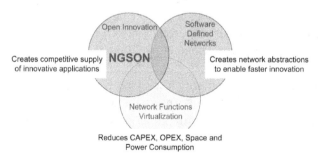

Figure 1   NGSON Interfacing SDN and NFV(adapted from ETSI NFV White Paper)

3. Know how to select best fit routes depending on a number of rules and policies
4. Preserve once gained knowledge

Any form of manual upfront provisioning or commissioning these data into a database would not meet the objectives of being truly context-aware and dynamically adaptive. NGSON needs to be able make decisions on the fly based on a set of rules/policies and previously gained knowledge.

An ontology with a descriptive logic reasoner is able to provide this functionality.

The following standards and open platforms are available to support this NGSON ontology:

1. OWL2 - An ontology descriptive language standardized by the W3C consortium, based on XML syntax.
2. Protégé - An open software platform for designing ontologies in OWL2, also providing a descriptive logic based reasoner released by Stanford University.
3. Protégé - An Descriptive Ontology for Linguistic and Cognitive Engi-neeringAn upper level ontology released by the Institute of Cognitive Science and Technology of the Italian National Research Council

2 Service Composition Ecosystem

2.1 Similarities between Service Composition, Workflows and Business Processes

Most networking services may be considered as self-contained, self-describing, modular applications that can be published, located, and invoked across a network. However, the ability to efficiently and effectively select and integrate inter-organizational and heterogeneous services at runtime is an important step towards the development of service applications. In particular, if no single service can satisfy the functionality required by the user, there should be a possibility to combine existing services together in order to fulfill the request.

Despite previous efforts, the service composition still is a highly complex task, and it is often already beyond the human capability to deal with the whole process manually. The complexity, in general, comes from the following sources. First, the number of services available increased significantly during the recent years, and one can expect to have a huge service repository to be searched. Second, services can be created and updated on the fly, thus the

NGSON Framework Diagram

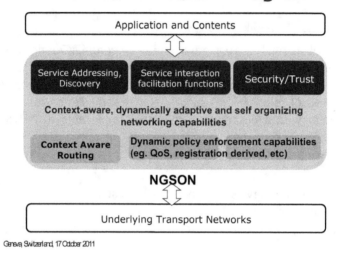

Figure 2   NGSON Framework

composition system needs to detect the updating at runtime and the decision should be made based on the up to date information. Third, services can be developed by different organizations, which use different concept models to describe the services, however, there does not exist a unique language to define and evaluate the services in an identical means. Therefore, building composite services with an automated or semiautomated tool is critical. To that end, several methods for this purpose have been proposed. In particular, most researches conducted fall in the realm of workflow composition or AI planning. For the former, one can argue that, in many ways, a composite service is similar to a workflow [2]. The definition of a composite service includes a set of atomic services together with the control and data flow among the services. Similarly, a workflow has to specify the flow of work items. The current achievements on flexible workflow, automatic process adaptation and cross-enterprise integration provide the means for automated services composition as well. In addition, the dynamic workflow methods provide the means to bind the abstract nodes with the concrete resources or services automatically.

A similar scenario can be observed in the modeling of business processes, where the Object Management Group has released a Business Process Model and Notation standard that uses workflow notational patterns to model business processes. There are also a few competing business process modeling languages available [3].

2.2 Using Deontic Logic to address drawbacks in current modeling schemes

A drawback of many current modeling schemes is that modalities are only implicitly expressed through the structure of the process flow.

All activities (service executions) are implicitly obligatory, i.e. mandatory, and whenever something should be permissible, i.e. optional, a branching node is used to split the modeling flow to offer the possibility to execute the activity or perhaps wait for an event or do nothing. This implies that the decision whether to execute one or more activities is described within that branching node, therefore the separation of decision and execution requires additional modeling elements and a comprehensive understanding of the entire process to identify obligatory and permissible activities.

Applying deontic logic can help here. While in classical logic, statements have values of either true or false, in deontic logic statements can be qualified by modalities. Modal logic began with Aristotle's analysis of statements containing the words "necessary" and "possible". These are but two of a wide range of modal connectives, or modalities that are abundant in natural and technical languages. Briefly, a modality is any word or phrase that can be applied to a given statement S to create a new statement that makes an assertion about the mode of truth of S: about when, where or how S is true, or about the circumstances under which S may be true.

Von Wright divided modal concepts into three main groups:

1. Alethic modes or modes of truth (necessary, possible and contingent)
2. Epistemic modes or modes of knowing (verified, undecided, and falsified)
3. Deontic modes or modes of obligation (obligatory, permitted, and forbidden)

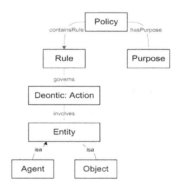

Figure 3 Top Level of Service Composition Policy Ontology

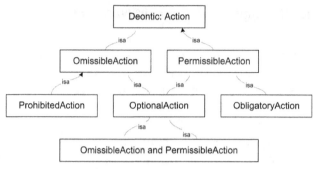

Figure 4   Deontic actions

Deontic Logic, therefore, is sometimes characterized as a form of modal logic describing the logic of prohibitions, permissions and obligations or the logic of ideal (according to certain norms) versus actual behavior.

Using W3C OWL2 Service Composition policies can be expressed in an ontology like:

'Modalizing' the ontology classes the reasoner will be able to deduct further service composition rules by applying Deontic Logic methods:

3 Interfacing with NFV and SDN

OWL2 ontologies may be augmented by forms of Rule Interchange Format, a further W3C standard. Rules in RIF provide the necessary functionality to specify procedural attachments in an OWL2 ontology, i.e. programs written in a procedural language that can be invoked by rules. These procedural programs may provide the necessary interfacing algorithms and code for communication with NFV and SDN functions.

4 Outlook

After compiling a current overview on available knowledge and semantic representation techniques as part of a Workshop on Software Defined, Context Aware and Dynamically Adaptive Services Overlay Networks (EcoSDSON) an attempt to provide a first, yet incomplete, OWL2 ontology in XML/RDF syntax should be undertaken.

Future research should also attempt to include temporal aspects into service composition design to accommodate for time relevant and dependent changes from a service as well as network related view.

References

[1] Jinghai Rao and Xiaomeng Su, "A Survey of Automated Web Service Composition Methods", Norwegian University of Science and Technology, Department of Computer and Information Science, N-7491, Trondheim, Norway, {jinghai,xiaomeng}@idi.ntnu.no

[2] F. Casati, M. Sayal, and M.-C. Shan. "Developing e-services for composing eservices." In Proceedings of 13th International Conference on Advanced Information Systems Engineering(CAiSE), Interlaken, Switzerland, June 2001. Springer Verlag.

[3] C Natschläger-Carpella, "Extending BPMN with Deontic Logic", Dissertation at Johannes Kepler Universität Linz, Austria, 2012

[4] "OWL 2 Web Ontology Language Structural Specification and Functional-Style Syntax (Second Edition)." W3C Recommendation, 11 December 2012. Available at: http://www.w3.org/TR/owl-syntax/

[5] "OWL 2 Web Ontology Language Direct Semantics (Second Edition)." W3C Recommendation, 11 December 2012. Available at: http://www.w3.org/TR/owl2-direct-semantics/.

[6] "OWL 2 Web Ontology Language Profiles (Second Edition)." W3C Recommendation, 11 December 2012. Available at: http://www.w3.org/TR/owl2-profiles/#Feature_Overview_3

[7] "RIF Framework for Logic Dialects (Second Edition)." W3C Recommendation, 5 February 2013. Available at: http://www.w3.org/TR/rif-fld/#Semantic _Framework.

[8] "RIF Core Dialect (Second Edition)." W3C Recommendation 5 February, 2013. Available at: http://www.w3.org/TR/rif-core/.

[9] C. Masolo, S. Borgo, A. Gangemi, "DOLCE : a Descriptive Ontology for Linguistic and Cognitive Engineering". Technical report, Institute of Cognitive Science and Technology, Italian National Research Council, 2003

[10] Handbook of Philosophical Logic, Vol II: Extensions of Classical Logic, Kluwer Academic Publishers Group, Dordrecht, Holland, 1984

Biography

Reinhard Schrage is an independent Telecommunications Consultant. He received his MSc degree in Mathematics and Computer Science at the University of Hannover in 1980. He has held several engineering and managerial posts at cutting edge technology providing enterprises and telecommunication incumbents. His assignments ranged from designing and providing customer tailored software for Government departments to responsible network planning for a globally operating financial services network. Now based in Germany he has studied in the USA, lived in New Zealand, the UK, Canada, Spain and Switzerland. He is a contributor to the IEEE P1903 'Next Generation Service Overlay Networks' Workgroup.

Author Index

Keywords Index

361